101 Tips for a Successful Automation Career

101 Tips for a Successful Automation Career

Greg McMillan and Hunter Vegas

Notice

The information presented in this publication is for the general education of the reader. Because neither the author nor the publisher has any control over the use of the information by the reader, both the author and the publisher disclaim any and all liability of any kind arising out of such use. The reader is expected to exercise sound professional judgment in using any of the information presented in a particular application.

Additionally, neither the author nor the publisher has investigated or considered the effect of any patents on the ability of the reader to use any of the information in a particular application. The reader is responsible for reviewing any possible patents that may affect any particular use of the information presented.

Any references to commercial products in the work are cited as examples only. Neither the author nor the publisher endorses any referenced commercial product. Any trademarks or tradenames referenced belong to the respective owner of the mark or name. Neither the author nor the publisher makes any representation regarding the availability of any referenced commercial product at any time. The manufacturer's instructions on use of any commercial product must be followed at all times, even if in conflict with the information in this publication.

Printed in the United States of America.
10 9 8 7 6 5 4 3 2

ISBN: 978-1-937560-50-8

ISA
67 Alexander Drive
P.O. Box 12277
Research Triangle Park, NC 27709

Library of Congress Cataloging-in-Publication Data in process.

Acknowledgments

There were many Monsanto and Solutia Directors and Senior Fellows who played a key role technically and personally in my journey as an automation professional. Foremost among those who shaped my career was the late Dr. Jim Fair, Professor Emeritus at the University of Texas, the creator and first director of the Engineering Technology (ET) Department. Being part of the ET Department allowed access to the best technical minds in the world, along with the freedom to explore and the encouragement to publish. My second ET director, Dr. Stan Proctor, shared his extensive knowledge in modeling and control and provided direction and support. The appreciation of my independence and humor by my last ET director, Dr. Adrian Hutton, another Texan, enabled me to publish many funny and creative articles and books. The knowledge I inherited from the late Vernon Trevathan and gained from conversations with Senior Fellows at Solutia and Monsanto, including Dr. Henry Chien, Dr. Bob Otto, Dr. Terry Tolliver, and Stan Weiner, has been of inestimable value.

When I retired from Solutia, I was fortunate to have the opportunity to become a consultant to Emerson Process Management at their headquarters in Austin in 2003. Here I have had access to some of the best minds on the supplier side, including James Beall, Terry Blevins, Mark Coughran, Lou Heavner, Mike Ott, and Willy Wojsznis. Jim Cahill has helped me to effectively use social media. The support by Mark Nixon, manager of DeltaV R&D, and Grant Wilson, vice president of DeltaV technology of my endeavors and the ISA Mentor Program has made this book and six other books since 2003 possible.

The participation and enthusiasm of the protégés in the ISA Mentor Program has inspired this book. The questions and answers posted on the ISA Interchange site gave us and the whole automation community an opportunity to step back and understand what type of knowledge is needed. Both Hunter and I thank the 2012 protégés Mohammed Al-Khalifah, Bahtiar Abu Bakar, Flávio Briguente, Brian Hartman, Luis Icasatti, Danaca Jordan, Madhawa Somasiri, Héctor Torres, and Bill Thomas.

The cover art, contributed by cartoonist Ted Williams (who also does the art for my humorous ISA books and my monthly *Control Talk* column), and Lucy Pless at Duke University, innovatively conveys the human dimension of the individual's journey. The extraordinary effort by Susan Colwell, manager of Publications Development, ISA, on evenings and weekends has enabled the ISA Mentor Program to be a success and this book to become a reality. Finally, the patience and understanding of my wife, Carol, have made this book and the sharing of what I have learned a part of our life.

~ Greg McMillan

Unlike Greg, I am new to the author game, and this is my first book (and acknowledgment). Writing this book has been quite a challenge, and there are several people whom I want to thank for helping make it possible.

I would like to start by thanking my first mentors—Clyde Barre, Henry Hecht, and Wedney Manuel. These three gentlemen showed extreme patience, took a completely green new hire under their collective wing, and taught me more than I could ever say about the field of automation and controls. I am forever indebted to them and thankful for their tutelage. I never could have become the engineer I am today without their guidance.

Thanks also to my brother, Michael Vegas, my colleague, Kevin Trantham, the ISA Publications staff, and my wife Carolyn for reading (and re-reading and reading again!) multiple iterations of these tips as they transitioned from badly handwritten scribbles on engineering graph paper to the version you see today. Their combined contributions did much to improve the final work product.

...and, finally, many thanks to my wife for letting me take over our kitchen table for these last three months as I labored through this book. I am happy to report that we don't have to eat outside anymore!

~ Hunter Vegas

Contents

About the Authors

Gregory K. McMillan

Greg is a retired Senior Fellow from Solutia/Monsanto and an ISA Fellow. At present, he contracts with Emerson DeltaV R&D via CDI Process & Industrial in Austin and consults for MYNAH Simulation Technologies in Saint Louis. Greg received the ISA Kermit Fischer Environmental Award for pH control in 1991, received the *Control* magazine Engineer of the Year Award for the Process Industry in 1994, was inducted into the *Control* magazine Process Automation Hall of Fame in 2001, was honored by *InTech* magazine in 2003 as one of the most influential innovators in automation, and received the ISA Life Achievement Award in 2010. He is the author of numerous books on process control, his most recent being *Essentials of Modern Measurements and Final Elements for the Process Industry, Advanced Temperature Measurement and Control, Second Edition,* and *Good Tuning: A Pocket Guide, Third Edition.* Greg has been the monthly "Control Talk" columnist for *Control* magazine since 2002, posts a "Control Talk Blog" each Thursday on the *Control Global* website, and posts tips and questions and answers for the ISA Mentor Program participants each Friday on the *ISA Interchange* website.

P. Hunter Vegas, PE

Hunter Vegas was born in Bay St. Louis, Mississippi, and received his BSEE degree from Tulane University in 1986. Upon graduating, he joined Babcock and Wilcox, Naval Nuclear Fuel Division in Lynchburg, Virginia, where his primary job responsibilities included robotic design and construction and advanced computer control. In 1987, he began working for American Cyanamid (now Cytec Industries) as an instrument engineer. In the ensuing 12 years his job titles included Instrument Engineer, Production Engineer, Instrumentation Group Leader, Principal Automation Engineer, and Unit Production Manager. In 1999, he joined a specially-formed group to develop next generation manufacturing equipment for a division of Bristol-Myers Squibb. In 2001, he joined Avid Solutions, Inc. as an engineering manager and lead project engineer, where he works today. He lives in Kernersville, North Carolina.

Hunter holds Louisiana and North Carolina Professional Engineering licenses in Electrical and Controls System Engineering, a North Carolina Unlimited Electrical Contractor's License, and an MBA from Wake Forest University. He has executed nearly 2000 instrumentation and control projects over his career, with budgets ranging from a few thousand to millions of dollars. He is proficient in field instrumentation sizing and selection, safety interlock design, electrical design, advanced control strategy, and numerous control system hardware and software platforms.

Preface

This book was inspired by what the authors have learned from the ISA Mentor Program. As guides for ten (10) talented and enthusiastic individuals, we have developed an appreciation of how formidable is the challenge faced by new engineers. Success in the automation profession depends on each engineer's being able to deal with largely undocumented user knowledge, an incredibly diverse range of applications, vast quantities of un-analyzed data, and expertise increasingly residing outside of the plant. In a creative, concise format, this book seeks to provide perspective, understanding, direction, and guidance on what has been, and will be, important for an engineer's advancement in an automation career. Growth in skills and knowledge is important to your company, to this profession, and to you in terms of promotion and marketability.

This book captures 101 of the most important ideas learned through more than 60 years of application experience by the authors. To increase effectiveness, each Tip includes a concept to extend applicability, critical details and watch-outs to prevent failures, a key insight to increase understanding, and a rule of thumb to provide a concise straightforward step. The book provides extensive practical information not found in the literature. Generalization in the concept, insight, and rule thumb is offered to enable you, the reader, to see the commonality in applications and build on knowledge gained.

The authors' backgrounds, skills, and knowledge complement each other to provide a more complete picture. Hunter's emphasis is on successful automation project execution, whereas Greg's focus is on benefiting from the knowledge of experts and resources and using the best technology for measurements, valves, modeling, and control.

Projects, in general, and process control improvement, in particular, are team efforts that depend, in large part, on the human factors of inspiration, motivation, and prioritization. In addition to technical expertise, the Tips offer ways to increase your skills and performance in interpersonal relationships, the shaping of attitudes, and the how and why of purpose-driven conversation. The book is designed to help you to achieve success in a given task and to get the recognition deserved, which will open the door to career advancement and expanded access to resources for future opportunities.

The book's cover captures the essence of the journey of the individual in a career. Each tip is a step to a higher level of achievement. The individual can see a door to openness, freedom, and perspective at the top of the automation profession for attaining the highest level of plant and personal benefits.

Tip #1: Always Ask "Why?" – And Never Stop Learning

Hunter Vegas (Philosophical)

Both Greg and I thought this tip was the perfect place to start a book like this. If we had to choose one trait above all others that places an engineer on the path to success, it would be a constant quest for knowledge. We have each spent our lifetimes in pursuit of this goal, and as our careers slowly fade toward retirement, we decided to pass some of our hard-learned lessons on to the next generation. Hopefully the next group of engineers will have the wisdom to learn the easy way rather than blundering their way up the learning curve as we have!

Greg and I are an interesting pair. While he has a lot of plant experience, he tends to concentrate more on process and dynamic principles and technology. I, too, have spent a great deal of time in plants but have been doing project execution and management my entire career. Our life experiences are very different, yet complementary, and somewhere along the way we decided we should write a book together. This is Greg's umpteenth book and my first, and as you can see, our writing styles are quite different. The first half of the book is written by me, and the second half by Greg. My tips will tend to be shorter and a little lighter technically. Greg's tips will tend be more involved, but provide some extremely useful information. Hopefully between the two of us, you will learn a great deal of our hard-won knowledge by the end of these pages. All we ask in return is that as you grow in knowledge, you will take the time to teach the NEXT generation behind you.

And without further ado let us begin…

Concept: Knowledge is power, and you should never stop learning. Seek to know and understand everything. The most valuable automation professionals know hundreds of processes, are well versed in electrical, mechanical, and chemical engineering, and have a breadth of knowledge in multiple industries. If an engineer is not learning something new every day, he or she is falling behind.

Details: This general concept will apply throughout this book. Whenever you are faced with a new process or industry, you should become a sponge. Listen, and listen some more, and stop listening only to ask more questions. Always ask "why"… even if you are embarrassed and think everyone else in the room understands but you. (Most of the people in the room probably do not understand either!) Constantly read technical publications, attend ISA and other training seminars, and listen to your colleagues.

Watch-Outs: Evaluate your source of knowledge carefully. A person who truly understands a subject will not mind explaining it in a different way until you understand it. A person who is PRETENDING to know will tend to be dismissive and get angry. Such self-proclaimed "experts" may not know nearly as much as they might claim. Also, beware of technical articles that have a sales and/or marketing spin to them. (See next tip.)

Exceptions: None.

Insight: Some of the most useful information comes from the operators who work in the plant every day. They may not fully understand the technical details, but their wealth of knowledge of the day-to-day problems and the dynamics of their plant can be invaluable to you. Building a strong relationship with the operators should be a top priority. If you watch their backs, they will watch yours.

Rule of Thumb: The more knowledge you master, the more valuable you become. Seek to know EVERYTHING, and never stop. Even experts in the automation field find themselves learning new things every day.

> **"If we had to choose one trait above all others that places an engineer on the path to success, it would be a constant quest for knowledge."**

Tip #2: Always Note the Author of a Technical Article

Hunter Vegas (Philosophical)

Of all of the engineering fields, I have to think that our field of automation has one of the fastest rates of change. Every day new processes, new technologies, new instruments, and new control techniques invade the market, and an automation engineer must constantly strive to stay abreast of the latest offerings or face becoming obsolete. To stay current, I find myself constantly reading technical articles so I can quickly evaluate the benefits and weaknesses of the latest entries and determine whether they might be useful to my company or my clients. Of course the sales and marketing departments of the various vendors know this, and they spend a lot of effort publishing articles to "help" engineers select their product over their competitor's.

Concept: Staying technically current in the field of automation is a never-ending task. Attending ISA meetings and conferences, vendor expos, and industry group conventions is a start, but your best solution is maintaining a steady diet of articles to keep abreast of the latest trends and technology. While many articles provide excellent information on a wide variety of topics, many others are written with a particular product bias or marketing angle. The author's description, position, or byline will often alert you as to which type of article you are reading.

Details: Sales and marketing departments often employ people whose sole purpose is to write articles that appear technical but are in fact specifically written to entice engineers to specify or purchase their products. Before you read any article, skip ahead to the author information at the end and determine who wrote the piece. Please note that this tip is NOT a global slight against technical articles written by vendor personnel. Many vendors employ the top experts in a particular field to design and develop their products. These people often write informative and unbiased technical publications that are invaluable sources of information. However, knowing the background of the author before reading a piece can help you be on the lookout for misleading statements or positions that seem to favor one product or technology to the exclusion of others.

Watch-Outs: Be particularly wary of technical publications/magazines that are sponsored by a single vendor. These publications rarely allow any disparaging comments about their product and will rarely mention a competing technology except to explain how their product is vastly superior. While these publications can be a good source of information about a single system, they tend to limit their focus to the company's product line and ignore all others.

Insight: Look for job titles such as "Sales Director," "Marketing Manager," "Product Development Manager," or the "President" or "Vice President" of a particular vendor. Such titles are a strong clue that the true purpose of the article might be more sales related than technical.

Rule of Thumb: Just because an article is written by the sales/marketing department does not mean it should automatically be ignored. However, if the piece IS written by the sales/marketing department, be on the lookout for a biased view of the product. Does the article mention a particular brand or technology exclusively? Does it offer pros and cons of the subject, or does it only mention advantages and benefits and never mention any negative aspects. These questions can help an engineer quickly determine if the article in question was sponsored by the marketing department.

> **"... your best solution is maintaining a steady diet of articles to keep abreast of the latest trends and technology."**

Tip #3: Pain Is Instructive

Hunter Vegas (Philosophical)

I must confess that this is one of my favorite lines because it applies to practically everything from managing your boss, to dealing with co-workers, to raising children. The concept is simple—humans (and most animals) learn to avoid "pain." (The word "pain" is used here in a very broad sense.) If we burn our fingers, we draw them back quickly. If we fail to get a pay raise for doing poor work, we tend to step up our efforts. If a fly is buzzing in our face, we swat at it. But what do you do if something is bothering you that you CANNOT resolve by your own actions? How do you swat the fly if your hands are tied?

Concept: Everybody encounters problems that are horribly frustrating, especially if we are dependent upon others to resolve them. Perhaps a co-worker is lazy and is not pulling his weight, or your child refuses to clean up his room, or another department keeps promising to execute a task but never gets around to it. The answer to all of these is to understand the fundamental truth—that pain is instructive—and then apply that concept. Let us be clear and state that this tip is NOT advocating threats or violence...as tempting as that may be! The trick is to figure a (non-violent) way of transferring the "pain" you are feeling to the person who can resolve the problem. If that can be accomplished, the other person has much greater incentive to work on the problem and make it go away.

Details: This idea is better explained by example. Consider the following:

- You are responsible for making sure all of the time sheets are entered by the entire department every week. Several people routinely fail to enter them on time and require constant badgering on a weekly basis. Rather than fight the battle, simply tell Accounting to not issue any paychecks to people who failed to enter their time sheets. After a check cycle or two, the problem will resolve itself.
- Your young child never washes out the sink after brushing his teeth and leaves the mess for you to clean up. Assign the chore of cleaning the bathroom to him so that he has to clean up that mess along with the rest of the bathroom every day. You will suddenly find him making the effort to keep the bathroom much more tidy so he has less work to do.
- You are on a softball team. Some players go to the batting cage every week before the game and have good batting averages. Others never practice and just show up for the game and bat poorly. You have tried to encourage the weak batters to do a bit of practice, but they refuse to be bothered. Rather than fight them, just list the batting averages sorted best to worst and post them in the dugout before each game. Once the weak guys start getting kidded for being at the bottom of the list, the incentive for moving up will become much greater.
- Your company laptop has problems, but the IT guy dismisses it as a non-issue and will "get to it when he can." Rather than argue, just offer to swap your laptop for his, since in his words "the problem is minor." If HE has to deal with a broken laptop on a daily basis, he will be much more inclined to make time to fix it.

Watch-Outs: Many people's first reaction is "to go to the boss" when they cannot easily resolve a problem. While this can be effective in the short term, it tends to put people off and causes rifts in relationships down the road. Avoid this if possible.

Avoid the tendency to be vindictive for past slights, and make sure that the link between cause and effect is clear to the person. You want the problem to be the cause of the discomfort ... you DON'T want to be the cause.

Exceptions: Just to be clear (again): this tip is NOT advocating workplace violence. As unhappy as you may be with a co-worker, you are not allowed to threaten him or whack him over the head!

Insight: Human relationships and interactions are not generally a strong suit for engineers. Rather than battle a problem, just turn it around so that the person who can resolve the problem is as vexed by it as you are. In a short time, they will WANT to fix it just to make it go away! With a bit of practice, you will find this can be done rather subtly, and people will marvel at your ability to get things done.

Also realize that just as pain can be instructive, "warm fuzzies" can be encouraging. Take the time to compliment those you work with when they do a good job. Everybody likes to get recognized.

Rule of Thumb: One of the more frustrating problems is the kind that can only be fixed by someone else, and they will not do it. In such a case, find a way to have that person feel the same frustration as you, and they will be much more likely to get it resolved.

> **"Take the time to compliment those you work with when they do a good job."**

Tip #4: Never Lie

Hunter Vegas (Philosophical)

I have managed hundreds of projects and engineers over my career, and one of the most important concepts I stress to my team is to tell me the complete truth—NOT what I want to hear. If a project is running behind schedule, I need to know NOW so I can adjust and possibly recover. If I am told that everything is fine and on schedule throughout a project, only to find problems at the end, then I have no option but to fail.

However, this concept goes far beyond project status reports. For example, overstating experience and skill sets on resumes may be common practice, but as a person who routinely reads resumes and interviews engineering candidates, I can unequivocally say that the quickest way to lose an interview opportunity is to overstate or lie on your resume. If I cannot trust you to truthfully fill out a resume, how can I possibly trust you to design control systems with people's lives at stake? Obviously, listing your work experience is important, but do not take credit for things you did not do and do not claim experience or knowledge you do not have.

If I had to pick one aspect of my personality that has helped advance my career it would be my reputation as a "straight shooter." If you ask me a question, I am going to answer it to the best of my ability, and if I do not know the answer, I will say so. Telling the truth breeds trust, and it is that trusting relationship with coworkers and clients that has served me well throughout my career.

Concept: This concept is straightforward. Do not lie, and do not tell people "what they want to hear" just to avoid conflict. People generally give a new acquaintance or business colleague the benefit of the doubt when they first meet them and assume they are a truthful person. However, once the first lie or half-truth is told, everything you say may be called into question. A reputation as a liar can stick with you for a lifetime.

Details: Work very hard to establish a high level of trust with your co-workers or clients. Do not commit to goals you cannot achieve, but always deliver what you promise. If the project is not on track, tell your co-workers or client (immediately) and seek advice on the best way to resolve it. Your word should be your bond. If you know the answer, provide it. If you do not know the answer, then saying "I don't know, but I will find out" is perfectly acceptable.

Watch-Outs: New graduates and inexperienced engineers tend to "pad" their resumes, hoping to get their foot in the door. The fact is that NO new graduate has much worthwhile experience, so the hiring firm has pretty low expectations in that respect. If a candidate declares himself an expert at programming because he wrote a handful of code or claims extensive experience when he has none, not only will he not get hired, he probably will not even get a shot at the first interview.

One perspective candidate submitted a resume via email to my company. One of the hiring managers happened to open the document up with "Track Changes" enabled, which highlighted all of the recent edits to the resume. Apparently the candidate had a degree in

Chemical Engineering, but decided to change it to "Electrical Engineering" since he assumed my company was more interested in that major. Ironically, we hire engineers with either of those degrees, but needless to say we did NOT hire him!

Exceptions: None.

Rule of Thumb: Tell the truth—always. A reputation as a liar can stick with you for a lifetime.

"Telling the truth breeds trust..."

Tip #5: Admit Your Errors

Hunter Vegas (Philosophical)

This tip is really an extension of the previous "Never Lie" tip. Every engineer in the history of the world has made mistakes. In fact, the best engineers are the ones who have made LOTS of mistakes—and learned from them. As mentioned previously, pain is instructive, and a painful mistake is a wonderful way to ensure that you will not make the same mistake again. However, if you never admit it happened, not only will you fail to learn, but others will not learn either.

Concept: Every engineer screws up. It is not a question of if, but when it will happen. Do not make a bad situation worse by failing to admit it.

Details: If you screw up, admit it early and publicly. Your co-workers or clients will think much better of you for admitting your fault rather than choosing to blame others or making excuses. Unfortunately, some individuals in engineering are too proud to admit their errors, and so they play the "blame game." Someone else is always responsible, or they make excuses to explain why the problem "really wasn't their fault." When a person acts this way, their colleagues catch on quickly and may either refuse to work with them or set them up to fail in a public way. Either way, the final result is always dramatically worse than if they had just admitted the mistake right up front.

Just as important as admitting your mistake is learning from it. Take time to understand what went wrong and make changes to ensure that it will never happen again. Better yet, go one step further and tell others about your mistake so they can avoid it themselves.

Watch-Outs: Be wary of co-workers and bosses who try to blame YOU for their mistakes. You will learn to spot them quickly. In such a situation, protect yourself with e-mails and paper trails.

Exceptions: As odd as it may seem, you can be TOO quick to admit a mistake. In the heat of a start-up (for example), a client can jump to conclusions and decide something is "wrong" when it really is correct. When a problem comes up, before you acknowledge that you made a mistake, take a moment to investigate it and determine whether there really is a problem. Then, if you have made an error, admit it and move on.

Insight: You will be amazed how a quick admission of error can defuse a potentially bad scene. When a client or co-worker discovers a problem, it may be tempting to deny or minimize the problem or push it off on someone else. However, the best response is to immediately acknowledge the problem, admit the mistake, and start working toward a solution. Even if the problem is NOT your fault, avoid the witch hunt and seek to resolve the problem first. Once the crisis is over, you can determine what went wrong, and everyone involved can learn from the error.

Rule of Thumb: When (not if) you screw up, admit it and move on. Admit your mistake and set about correcting it. You never actually fail until you give up.

In fact, the best engineers are the ones who have made LOTS of mistakes — and learned from them.

Tip #6: Change for the Sake of Change Is Not Always a Good Thing, but "We've Always Done It That Way" Is a Poor Reason to Avoid Trying New Things

Hunter Vegas (Philosophical)

I am a risk-averse engineer. My clients do NOT like surprises, and they pay me a decent salary to make sure that their automation projects go in as painlessly as possible. Therefore, I do not apply new technologies unless I know they work and I am content to let others debug the latest software revision before I upgrade to it. Despite all that, I absolutely HATE the expression "....but we have always done it that way." I have no problem with "We do it that way because we tried x and y happened" or "We considered that technology but chose this other method because...", but when a company just refuses to try a technology because it is "different" it drives me crazy.

Concept: Sticking with tried and true techniques that work and are risk-free certainly makes sense. Living on the "bleeding edge" of technology is painful. However, that is no excuse for failing to investigate and try new things.

Details: Automation professionals are trained to avoid risk. That is a good thing because trialing untested pieces of equipment or applying novel methods of safety shutdown when designing a control system can have severe consequences if things go awry. However, that is NOT a reason to avoid making any changes at all. Obviously, the best solution is somewhere in the middle.

Some engineers take great delight in getting the latest version of software or specifying the latest technology product. However, companies often release software versions after minimal testing and rely on their customers to "beta test" their product for them. Rather than debugging code for these firms, the wisest course of action is to lag behind by a software revision or at least wait for Revision X.1 to be released, which fixes the bulk of the bugs from Revision X.0. Similarly, it can be best to delay hopping on the "technology du jour" bandwagon, because despite what the marketing circulars say, all technologies have pros and cons, and no one product or technology is the panacea for every woe.

However, decision-makers at some plants refuse to change ANYTHING because "It has always been done that way, and that is the way we do it." This ostrich mentality hamstrings a plant's future growth and profitability. Keeping abreast of new products and technologies as they are offered and taking advantage of them when it makes sense keeps a plant competitive. Look for opportunities to try new equipment or software on non-critical systems, where the financial and operational risks are low. Occasionally vendors will allow you to try a particular instrument for free and only pay for it if it works in that application.

When making a change is appropriate, make sure that you build a compelling argument for the change, listing the risks and benefits. Nearly all change is going to cause some short-term discomfort. People need to understand the long-term benefits so they will stop fighting the change and will work toward a better future.

Watch-Outs: If a plant offers resistance to a new idea, do NOT immediately assume they are just afraid of change. Investigate what has been tried before, and find out exactly why it went wrong. The "old timers" can be an invaluable source of information. There may be some aspect of the process that will not allow the proposed technology to work, and a conversation with the right people could help you avoid an embarrassing failure.

Do not assume that everyone who went before you were idiots. Some might well have been, but dismissing all of the work that has been accomplished previously means recreating everything from scratch. That will invariably require a lot of time and money and will probably force you to re-learn the hard lessons already learned by your predecessors.

Insight: One way to avoid problems when considering making changes is to develop a network of automation engineers in a couple of plants, or ideally across a couple of industries. (ISA or other technical societies can be an excellent means of doing that.) When considering a new technology, ask around and find out how others have fared.

Rule of Thumb: Do not be scared to try new equipment or software but be wise in deciding when and where to try it. Realize that the first version of nearly every software product will be rife with bugs and problems. If possible, wait for the next revision release. Similarly, "Serial #1" hardware or equipment that has just been undergone an extreme re-design will likely have some flaws that will take a generation or two to rectify.

> **"Automation professionals are trained to avoid risk ... that is NOT a reason to avoid making any changes at all."**

Tip #7: If You Have to Tell Everyone How Smart You Are, You Probably Aren't

Hunter Vegas (Philosophical)

I have met many people who were infinitely smarter than I, and their intelligence was obvious to me within a few sentences of conversation. Conversely, I have met other people who were quick to inform me of their supposedly advanced intellect, and in most cases they were not very bright at all. If you have to tell people how smart you are, then you probably aren't; and if you ARE smart, everyone will know it regardless of what you say. Do not get hung up on showy displays of college degrees, awards, etc. because they may not be a good indication of a person's ability. I know a great many people who never received a degree of any kind, yet they are some of the most knowledgeable and respected engineers in the field. Similarly, I know too many people who have advanced college degrees and have awards and recognition plaques all over their office walls, yet they are incapable of doing the simplest engineering designs. Like most people, I have tremendous respect for a person who is obviously brilliant yet downplays it.

Concept: If a person is extremely intelligent, it will be obvious to everyone after a casual conversation. However, the people who brag about their own intelligence are rarely as smart as they want you to think they are. The most talented and brilliant people will often say the least and listen the most.

Details: We have all encountered the type of person who feels obligated to display every award or plaque they ever owned, highlight their superior intelligence at every opportunity, and talk down to "the little man" whom they consider to be beneath them. Ironically, people like that are often not very gifted at all, but carry on the show to make themselves look that way. Do not ever allow yourself to be counted in that number.

On the other hand, it is quite impressive to meet individuals who do not have anything on their wall, constantly downplay themselves and their accomplishments, and yet are true geniuses. These people ask for others' opinions and seek help from everybody, never put down others who are less educated or less knowledgeable, and are generally well liked and respected by all. Within moments of meeting and talking to a person like that, you know that he or she IS a genius, and yet they will never make mention of it. That is the person you want to emulate.

Never, EVER look down on a person because they lack the education or position that you have achieved. From the CEO to the lowest level employee, everyone knows something that you do not and they can often be valuable sources of information and new ideas. Treat them with respect, as you would want to be treated, and you will be amazed what they can teach you.

Watch-Outs: Such things as diplomas on the wall, awards prominently displayed on the shelves, early and frequent mentions of advanced degrees, insistence of being referred to as "Doctor." Such a person is hard to miss. Do not be one of them.

Exceptions: Occasionally you will encounter a person who acts like a "know-it-all" and actually is a know-it-all. These people are fairly rare, but they do exist. They typically have strong, aggressive personalities and are smart enough to adapt their personality style to the situation. Dealing with such a person can be trying at times, but at least they have intelligence to back their bravado and their knowledge can be extremely helpful when you are faced with a technical dilemma.

Insight: Even though many technicians and operators may lack higher education, they are usually extremely knowledgeable of the plant and its operation. Their knowledge of the process and its hidden interactions and problems as well as the effort it takes to keep the plant running is invaluable to an engineer. Foster a strong relationship with them and listen when they offer information. They can be an excellent audience for evaluating new automation ideas and will often tell you what problems or pitfalls you might encounter. What is more, once they realize that they are being heard, they will provide new ideas and suggestions that might never have occurred to you. Cultivate those relationships, and you will reap rewards throughout your career.

Rule of Thumb: The smartest person in the room rarely has to prove it—everyone knows who he or she is. Learn to be that person.

"The most talented and brilliant people will often say the least and listen the most."

Tip #8: Never Underestimate the Power of Politics and Emotion

Hunter Vegas (Philosophical)

The power of politics and of human emotion can be mind-boggling and utterly baffling to engineers who are taught throughout their lives to apply sound logical principles and facts to decision-making. I cannot begin to count the number of times when I have found myself desperately trying to apply logic to a situation where absolutely none could be applied. You cannot change the fact that politics and emotion are often intimately involved in a situation, but you can recognize that they exist and act accordingly.

Concept: For better or worse, engineers tend to be less emotionally driven than most. Therefore, they can get confused and blind-sided when people make a snap judgment based on feelings or when people choose a course of action dictated by some hidden political agenda rather than one based on the sound, logical principles that have been delivered for review. This can be most baffling to members of our profession.

Details: The fact is people often make their decisions based on emotion or politics, and the political or emotional angle often outweighs the logical argument. Unfortunately, most engineers do NOT think this way, and tend to assume others evaluate problems and information the same way that they do.

However, all is not lost if you recognize and accept this fact and adapt accordingly. If emotion and politics are afoot, then learn the rules and play the game! This is NOT an invitation to wade into corporate politics—life is too short—but this is a suggestion to study and understand the role of emotion and politics in decision-making so that their effects no longer appear as a surprise. Learn to separate the logical/technical aspect of problems from the political/emotional aspect so your effort is not wasted using the wrong skill set to resolve an issue. As ironic as it might sound, once the emotional or political angles of an issue are known, logic can be applied to resolve it. You need only apply a different set of rules. For instance if the decision becomes an emotional one, frame the arguments to cater to that mindset. If politics are driving a decision, seek to understand the source of those politics and act to convince the true power brokers/decision makers of your position.

The effects of politics or emotion often get much worse when there is an audience. People will defend a poor position to the death before they will retreat in front of their peers. Sometimes it is better to talk through disagreements one-on-one after the meeting and out of sight of others. Better yet, try not to let the situation develop to that point.

Watch-Outs: Despite the fact that most engineers consider themselves to be extremely logical, they can be emotional themselves. If you find yourself banging heads with a manager or co-worker, consider standing back and examining the situation as a disinterested third party (or discussing it with a disinterested third party). It may be that YOUR emotions are clouding the issue.

Exceptions: On rare occasions, people will ignore all of the politics and emotions swirling around an issue and will make a sound, logical decision. Oh how I wish this was the rule rather than the exception!

Insight: As an aside, you can eliminate a lot of conflict in your lifetime by simply avoiding discussing topics which are certain to create conflict in the first place. Many topics (religion, political candidates, favorite sports teams, etc.) can be very polarizing and are almost certain to cause problems. If both parties agree, there is really not much to discuss. If the parties disagree, then in all likelihood neither can say anything that will sway the other's position. Why fight the battle?

Rule of Thumb: Always be aware of the emotional and political angles of an issue. If either is present, then recognize that the rules have changed and adapt accordingly. Never try to apply logic where it does not belong.

> **"If emotion and politics are afoot, then learn the rules and play the game!"**

Tip #9: The O-ring Seal in an On/Off Actuator Can Be a Good Predictor of Its Reliability

Hunter Vegas (Technical)

Over the course of many years of plant experience, I have come to a simple conclusion with respect to selecting pneumatic on/off actuators. Despite innumerable glossy color sales brochures and sales presentations to the contrary, the failure of an on/off actuator can usually be attributed to one or more of three things. Two of these will make ANY pneumatic actuator fail: undersizing and poor air quality. The third is more subtle, yet it can be a surprisingly accurate predictor of how well an actuator will hold up in service.

Concept: Several different on/off actuator designs are available today. Some employ a scotch yoke mechanism, others use a rack and pinion, and there are other variations. While vendors will argue the pros and cons of one design versus the other, plant experience suggests that it is the design of the piston O-ring that is the best indicator of how well a particular actuator will hold up in service.

Details: An on/off actuator converts air pressure to a 90 degree turn movement that actuates a valve. Most actuator failures can be attributed to three things:

1) The actuator was undersized from the start. (See Greg's Tip #81 concerning this.)
2) Poor instrument air quality: If water and/or particulates are in the instrument air system, every positioner, solenoid, and pneumatic actuator in the plant will fail prematurely.
3) The piston O-ring fails.

If the actuator is properly sized and the air quality is good, the typical point of failure will almost always be the piston O-ring, which seals against the cylinder. Once this O-ring begins to wear, it will allow the air pressure to "blow by" the piston, robbing the actuator of torque. Eventually, the actuator will not stroke at all. The design of this O-ring is what usually determines how long an actuator will last in service.

A cheap design will employ a single round O-ring on the piston. Such a design works wonderfully when new, but quickly wears and begins leaking air. By contrast, a better design will employ multiple O-rings or a wide, flat ring around the circumference of the piston. Either of these designs will last much longer.

Watch Out: A poor O-ring design will make an actuator fail quickly, but also watch out for actuator limit switch covers that employ individual screws that are not captive in the cover. (In other words, the screws fall out when they are unscrewed rather than being held in the cover.) Captive screws may not seem like such a big deal until someone is on the 5th floor of an open structure pulling a cover to set a limit switch and the screw falls out and bounces off several vessels and pipes on the way to the ground below. At that point, the utility of captive screws in the cover becomes obvious!

Exceptions: Some designs use a shorter stroke and a diaphragm instead of a piston with an O-ring. This particular design is not susceptible to O-ring wear.

Insight: Find at least two acceptable actuator designs and get both vendors on your bid list. Having two sources helps to keep the pricing low and limiting the actuator types to only two cuts down on spare parts inventory.

Once a good actuator design is chosen, always oversize the actuator. If the instrument air quality is good, the actuators will provide years of maintenance-free service.

Rule of Thumb: Pick a good design, and size the actuator for at least one and half times the maximum torque required. (Note that actuators have different torque values at either end of the stroke so be sure to check both ends of the table when doing the sizing.)

"Once a good actuator design is chosen, always oversize the actuator."

Tip #10: Butterfly Valves and/or Ball Valves with Positioners Are NOT Control Valves!

Hunter Vegas (Technical)

One of the more heated arguments I had with a client involved his extreme desire to use line size butterfly valves for controlling his process. The entire concept of using control valves for the available pressure drop and for controllability at low flow rates was utterly lost on him. We went round and round until I finally got the client to install spool pieces at each valve location so we could easily replace the butterfly valves with a different valve if we encountered control problems in the future.

A lot of those spool pieces and butterfly valves got replaced.

Concept: Despite what the on/off valve sales person will tell you, an on/off valve with a positioner is NOT a control valve. Such an arrangement can have its uses, but you should understand the limitations of on-off valves and the differences between on-off valves and control valves.

Details: In an attempt to reduce cost, a young engineer may resort to using ball and/or butterfly valves with positioners to control a process. These valves CAN be used in throttling applications, but certain limitations can greatly impact their performance. An automation engineer must understand these limitations and specify this type of valve only where appropriate.

PROS:

1) This arrangement is much cheaper than a standard control valve.
2) A ball valve's flow characteristic can be "characterized" to provide a range of CVs and provide a more linear response if required. (Butterfly valves are not so easily modified.)
3) On/off valves tend to have less pressure drop when fully open.

CONS:

1) A butterfly valve's flow characteristic is essentially closed until about 40 degrees, then flow increases from 20% to 90+% of capacity between 40 and 80 degrees. Such a narrow range of control and poor turndown rarely suits most control valve applications.
2) The recovery factor for a butterfly or ball valve is generally poor compared to a control valve. Therefore, these valves tend to have more problems with cavitation and permanent pressure drop when they are throttling flow.
3) Precise, repeatable control is difficult to achieve – especially at low flow rates.

One application that IS well suited to a throttling on/off ball valve is a dribble application in batch raw material charging. The typical dribble valve uses a second solenoid to "lock" the valve position in a half-closed position, which slows the charge rate at the end of a charge and helps improve the accuracy of liquid charges. However, the dribble valve solenoid arrangement is prone to maintenance problems and the final charge rate and accuracy tend to drift as the actuator ages. For nearly the same money, a small positioner can be installed on a characterized ball valve and the unit can be continuously throttled at the end to provide very accurate charging. Typical control logic would look something like this:

- If remaining charge amount > 100 lbs, output = 100%
- If remaining charge is <=100 lbs, output = MAX (remaining charge/100, 25)

Such an algorithm charges at maximum speed until the end, then continuously throttles the valve down to 25% as the end of charge approaches. Finally, it trickles in the last few pounds to nail the total charge amount accurately.

Watch-Outs: Beware of using butterfly valves with class VI shutoff in any kind of a throttling application unless the valve is normally at least one-third open. These valves usually torque into the seat very hard and take a great deal of force to crack open from a fully closed position. If the valve is used to throttle in this regime, it will wear quickly and the flow rates will be wildly erratic.

Exceptions: If accurate control at low flow rates and cavitation is not a concern, then ball and butterfly valves can be successfully used in certain applications. However, you must understand the limitations of these valves types and not misapply them.

Insight: Premature valve failure usually occurs due to seat erosion under low flow/ high pressure drop conditions or under conditions where the valve is cavitating or flashing. True control valves can be designed to handle and/or mitigate these conditions. Throttling on/off valves are a poor choice for these applications.

Rule of Thumb: Carefully evaluate the application before selecting a valve type. Take time to understand the strengths and weaknesses of the various valve offerings, and choose the valve type that fits the application. Do not skimp on the design to save a few dollars; over the life of the valve, the wrong valve will cost much more than the original savings.

> **"Despite what the on/off valve sales person will tell you, an on/off valve with a positioner is NOT a control valve."**

Tip #11: The Good, the Bad, and the Ugly of Vortex Flowmeters

Hunter Vegas (Technical)

Every instrument is well suited for some applications and perfectly awful for others. These next few tips discuss the more common types of flowmeters and provide an insight into how they work, when they should be used, and when they should be avoided. I will begin with a brief description of how each meter works, and then discuss the pros and cons.

NOTE: Greg McMillan provides a good discussion of Coriolis meters later in the book so that type of meter is not included here.

Concept: Vortex meters can be an excellent choice for a large variety of applications. However, certain limitations associated with this type of meter can make it a poor choice in some applications.

Details: Have you ever watched a flag wave in the wind? It actually waves because the flag pole generates eddies (whirlpools, or vortices) on alternating sides of the pole, which move past the flag. The eddies are relatively high pressure areas that "push" the flag away and because they form on alternating sides of the pole, the flag weaves between them and flutters in the wind. The vortex meter works the same way. It employs some kind of vertical bar or "bluff body" that generates eddies (vortices) on alternate sides as the fluid flows past. Small sensors in (or behind) the bar detect the vortices and count them. The rate of vortex creation is chiefly dependent upon the rate of flow but is also dependent upon the viscosity of the fluid. (If the fluid is too viscous or the flow too low, the meter will not shed any vortices at all.) The meter converts the meter count into a fluid velocity and determines a volumetric flow rate by multiplying the fluid velocity by the cross sectional area of the meter.

Here is a quick list of the pros and cons of this type of meter:

PROS:

1) Works in gas, steam, and liquid applications.
2) Is insensitive to fluid conductivity.
3) It generally has a lower pressure drop than an orifice meter.
4) It is usually much cheaper than an orifice meter for line sizes smaller than 6" because it does not require impulse lines nor any special freeze protection beyond that of the existing pipe.

CONS:

1) Vortex meters require turbulent flow to operate and will cease to read as the fluid transitions from the turbulent flow regime to the transitional or laminar flow regime. This is called the "low flow cutoff" point of the meter, and flow rate measurement below this point is not possible. (See further information on this below.)

2) Vortex meters make effective (if unintended) start-up strainers. The vortex shedding bar (bluff body) across the meter is great for catching bolts, drink cans, oyster shells, and all kinds of other debris wandering down the line. When material gets caught on the body, the meter will either read inaccurately or not at all.
3) Some vortex meters use small ports to measure the vortices. These can plug with polymer or solids, keeping the meter from functioning. The design of the vortex measuring sensors is the chief difference between meters and the sensor design will allow some meters to work in certain applications where others will not.
4) High vibration or entrained solids can be problematic for vortex meters, which may detect and count solid particles or vibrations as if they are actually flow. (Note that most meters have a "noise band" or similar adjustment that can be set to reject these vibrations, but this makes the meter less sensitive and significantly increases the low flow cutoff point.)
5) Like an orifice meter, the vortex meter requires an upstream and downstream meter run to establish a good flow profile. This run should be no less than 15 diameters upstream and 5 diameters downstream, but most vendors like to see at least 25 diameters upstream and 10 diameters downstream.
6) A vortex meter measures volumetric flow—not mass flow. It can calculate a mass flow based on an assumed density, but if the fluid density changes, the reading will be in error.
7) Beyond 6" lines the economic advantage of vortex meters falls off.

Watch-Outs: The low flow cutoff is the chief limitation of a vortex meter, and it makes the meter unsuitable for any application where measuring low flows at high turndown is required. The low flow cutoff point is determined by the viscosity of the fluid and thus may vary with fluid temperature and composition. Highly viscous fluids cannot be measured with a vortex meter.

Exceptions: While most vortex meters are poorly suited for measuring liquids which tend to polymerize, some meters employ a proprietary sensor design that is not easily plugged and has a constant flow of liquid around the sensors to help keep them clean. Such units can work where others tend to fail.

Insight: For high temperature applications, be sure to specify a remote mounted transmitter head. The electronics do not like being cooked, and a locally mounted transmitter will not last long. Be on the lookout for centering rings if the meter is a wafer type. The pipe fitters tend to leave these in the box and just bolt the meter between two flanges. If the meter is not centered, it will be inaccurate.

Rule of Thumb: Vortex meters are a good choice for measuring the flow rate of any reasonably clean fluid where measurement at very low flows is not required. In line sizes of 6" or less, they are usually much cheaper than orifice plates, and they eliminate the need for impulse line heat tracing.

Tip #12: The Good, the Bad, and the Ugly of Magmeters

Hunter Vegas (Technical)

Early in my career, I was asked to specify a meter to measure the sideboiler bottom flow of an ammonia vaporizer. The sideboiler boiled off the ammonia for the process, and over time water would build up in the bottom of the sideboiler and had to be drained off through the meter. I had Operations pull a sample, and the conductivity of the ammonia/water mixture was very high. Given the high conductivity and the need to measure low flows, I considered a magmeter (magnetic flowmeter) to be a good choice for the application. Two weeks after the meter was installed, Operations called in a panic saying that the meter was reading zero even though liquid was obviously pouring through it. A quick investigation showed I had failed to consider the fact that if too much water was drained off, pure ammonia would go through the meter. Ammonia has extremely low conductivity, and at that point, the meter could not function and read zero. Since then, I always ask for the conductivity of the fluid during normal AND upset conditions!

Concept: Similar to the vortex meter, a magmeter can be an excellent choice for a large variety of applications, but it too has limitations. Understanding these limitations can help avoid a misapplication of this technology.

Details: When a conductor moves through a magnetic field, it generates a voltage. The higher the velocity of the conductor (with the magnetic field strength held constant), the higher the voltage generated. A magmeter uses this phenomenon to measure flow. In this case, the fluid is the conductor, and it flows through a non-conductive line sized tube that has a magnetic field passing from top to bottom. The meter has a pair of small electrodes (one on either side of the tube), which detect the resulting voltage and calculate a fluid velocity. The velocity times the cross sectional area of the meter provides a volumetric flow rate. Some meters include an additional electrode at the top and bottom of the tube to detect whether or not the pipe is full. (A half full pipe will read high because the calculation assumes that the pipe is full.)

Here is a quick list of the pros and cons of this type of meter:

PROS:

1) The meter is a full bore device and has practically no pressure drop.
2) The only metal contacting the fluid is the electrodes, which are very small. Therefore, the meter can be used to measure strong acids and caustics. Even if the electrodes must be made of an exotic metal (platinum, tantalum, etc.) the additional cost is not great.
3) The meter has no problems measuring acids, caustics, or conductive liquids with entrained solids. It can also work well for viscous fluids.
4) The meter effectively averages the flow profile of the pipe so it does not require the long, straight upstream and downstream meter runs that orifice or vortex meters require. Two to three diameters upstream and downstream are usually all that is required.

5) Recent improvements in electronics have made 2-wire magmeters available. (Older models required a separate source of 120VAC power.) The 2-wire devices are cheaper to install.
6) Magmeters do not have a low flow cutoff problem and will generally read as low as 1 foot/second.
7) If a remote transmitter is used, magmeters can handle very high temperatures and pressures.

CONS:
1) The meter only works on conductive fluids and will read zero if the fluid has no or very low conductivity. Most require at least 5 micromho, though some units can measure below that.
2) Cheaper magmeters use a PTFE or PFA liner with no additional reinforcement. When these meters are "steamed out," the tube can soften and if the pipe is then blocked in, the resulting vacuum in the line can collapse the liner and ruin the meter. A better meter uses a PTFE or PFA liner that has been reinforced by a wire mesh or frame to prevent this problem.
3) Ceramic magnet magmeters can easily handle higher temperatures and vacuum conditions; however, they can be prone to damage from thermal shock. If the fluid temperature is high, be sure to specify a remote mounted transmitter.
4) A magmeter will not work on a lined pipe unless ground rings are added between the flanges of the meter and the lined pipe. (These rings complete the circuit that allows the voltage to be generated.) Recognize that these rings will also touch the fluid and should utilize the proper material of construction.
5) Beware of tantalum electrodes (which are often used for strong acids). If these electrodes are exposed to air they will generate a non-conductive oxide coating which will keep the meter from operating immediately. Once they are again exposed to the acid, it will burn the coating off but this can take some time and the meter may not function at all during this time.
6) A magmeter measures volumetric flow—not mass flow. It can calculate a mass flow based on an assumed density but if the fluid density changes, the reading will be in error.

Watch-Outs: Beware of gravity flow measurements when using a magmeter. Unless the meter is properly located, partially empty pipe conditions will occur, and the meter may be inaccurate.

Exceptions: Always ask about the upset conditions that the meter might see. As mentioned previously, steam-outs can irreparably damage the meter and very low conductivity conditions can prevent the meter from reading at all.

Insight: Many magmeters employ a combination of AC and DC excitation on the coils to provide a means of detecting and compensating for coating of the sensing electrodes. While this may not make them impervious to coating conditions, it will allow the meter to continue to operate longer before a cleanout is required.

Rule of Thumb: A magmeter is an excellent choice for measuring the flow rate of a conductive liquid with reasonable accuracy. This meter is also well suited for measuring the flow rate of viscous liquids, acids, caustics, and slurries.

Tip #13: The Good, the Bad, and the Ugly of Differential Pressure Flow Devices

Hunter Vegas (Technical)

There have been whole books written about differential pressure (DP) flow devices (orifices, pitot tubes, flow nozzles, venturi meters, elbow taps, wedge meters, averaging pitot, etc.) I only have space to hit the highlights and suggest that you pick up any of several books on the subject if you want to know more about this type of meter.

Concept: Just like every other flowmeter, a differential pressure flowmeter has strengths and weaknesses. The DP meter is one of the oldest means of flow measurement, and many derivatives of the basic meter exist. This section will try to briefly provide the strengths and weaknesses of the offerings and alert you to the possible pitfalls of using this type of meter.

Details: Nearly all differential pressure flow devices use some version of Bernoulli's Law. Bernoulli's Law states that the energy of a fluid is composed of three types: static head, pressure, and velocity. If a fluid is forced to pass through a restricted area, then the velocity will increase in that restriction. Because the overall energy must remain the same, the pressure of the fluid will drop as the velocity increases. The pressure drop can be measured to calculate flow, but the pressure drop increases as the square of the flow, so if the flow doubles through a DP meter, the pressure drop will increase by FOUR times.

The differential pressure can be generated in a number of ways. An orifice plate is the most common, but flow nozzles, averaging pitot tubes, wedge meters, venturis, and many others can be used. It is even possible to measure the differential pressure across the inside and outside of an elbow to determine flow rate.

At a high level, here is a list of the pros and cons of this type of meter:

PROS:

1) Works in gas, steam, and liquid applications.
2) Is insensitive to fluid conductivity.
3) A well designed orifice installation can be extremely accurate, especially if the meter uses a honed meter run and temperature and pressure compensation. Such meters are often used for custody transfer applications.
4) Flow nozzles, pitot, and averaging pitot tube meters can generate very low permanent pressure drops. (Note that the turndown of these meters can also be quite limited.)
5) Averaging pitot arrays can accurately measure gas flow in odd shaped ducts with minimal meter runs.
6) If the line size is large (greater than 6"), the economics favor differential pressure meters over vortex meters. Pitot tubes, averaging pitot tubes, and several insertion type nozzles can be used in extremely large pipelines at relatively low cost.

7) Orifice meters do NOT have a low flow cutoff like vortex meters. If a high turndown is required, a high and low range transmitter can measure the DP across the same orifice plate to provide continuous flow measurement through a wide range of flow rates.
8) A segmented wedge meter can be combined with capillary seals to provide flow measurement of viscous, sticky fluids such as tar and sludge. This same arrangement can also handle some solids entrainment.
9) Because the DP transmitter is usually well removed from the process, DP flowmeters can be used in a wide range of temperatures and pressures.
10) Integral flow orifices can measure extremely low flow rates, well below those measurable by vortex meters.

CONS:
1) The square relationship between differential pressure and flow greatly limits the turndown of most DP meters. The pressure drop rises quickly as the flow increases.
2) The permanent pressure drop of most orifice plates is about two-thirds of the measured differential pressure. The energy cost can be significant over the life of the meter.
3) DP meters are sensitive to installation. The length of the meter run, the location of the transmitter, the method of running the impulse lines, and several other items can greatly impact their accuracy.
4) The impulse lines on liquid and steam meters usually require freeze protection. Impulse lines are also prone to plugging in many applications.
5) Like a vortex meter, the orifice meter requires an upstream and downstream meter run to establish a good flow profile. Such a run must usually be at least 25 diameters upstream and 10 diameters downstream. (Note that these distances depend upon the particular type of DP meter used.)
6) For lines smaller than 6", a vortex meter is usually a more economical choice if the low flow cutoff is not an issue.

Watch-Outs: The slightest bit of erosion on an orifice plate or pitot tube nozzle can have an enormous impact on accuracy. In addition, the sizing calculations are made for a very specific set of pressure, temperature, and density conditions. If these conditions can vary, the engineer should measure these parameters and adjust the flow rate calculation.

Exceptions: Each type of differential pressure flowmeter has a variety of pros and cons but the sheer number of types provides many options to the engineer. Few processes exist that cannot be handled by at least one version of the DP flowmeter.

Insight: Orifice type meters have been around for a long time and have been extremely well studied. Hundreds of constants and factors now exist, which can be included in the sizing calculations to make them highly accurate.

Rule of Thumb: Vortex meters have been gradually replacing orifice type meters, especially with the advent of the vortex meter made for steam applications. However, there are many applications where only a certain type of DP meter will suffice, and the DP meter (in its many variations) remains one of the most common flowmeters in the world today.

Tip #14: Capillary Seal Pitfalls

Hunter Vegas (Technical)

To master engineering design, you must master the art of trade-offs. An engineer is constantly balancing one criterion against another, gaining something here but giving up something else there. There will often be several factors to consider, all of which may counter or offset each other to varying degrees. Picking the right combination of features to suit the application can be challenging.

Specifying a capillary seal assembly is a perfect example of this.

Concept: Choosing the correct capillary seals for a particular transmitter installation seems like a minor thing, until you begin to understand the multitude of design decisions involved. Many an engineer has failed to grasp this and has gone through several meters until they got one that worked.

Details: Capillary seals are used to isolate a pressure or differential pressure transmitter from the process by transferring pressure from the process to a remote mounted transmitter. Some processes are prone to plugging of the impulse line, and the installation of a 2″ or 3″ seal in a full size line is much less likely to result in plugging than a typical ½″ piece of tubing would be. In addition, sanitary applications use a lot of capillary seals because they are easier to clean. A capillary seal consists of the seal itself (which is a flexible diaphragm), a piece of capillary tubing, and a standard pressure or DP transmitter, all carefully filled with a hydraulic fluid that has had all vapor removed. When pressure is applied to the seal, it is transmitted via the hydraulic fluid to the transmitter. Differential pressure transmitters will often, (but not always), have two seals, one on each side.

Here is a brief list of items that can cause an engineer serious problems:

- If two seals are installed on a differential transmitter, make the seals the same size and the capillaries the same length if possible. (This may require coiling up the unused length of capillary on one side.) The problem is that all hydraulic fluids expand with temperature, and the overall expansion is a function of volume. If the seals are the same size and the capillaries are the same length, the hydraulic expansion from one side will cancel the other, and the overall zero shift will be minimized. If one leg is longer or one seal is bigger, the hydraulic expansion will be greater on that side, and the zero shift can be significant.
- Seals with a bigger diaphragm are more sensitive and can measure lower pressures. However, bigger diaphragm seals have a larger volume and tend to show a larger zero shift due to process temperature changes. Smaller diaphragm seals have less volume and tend to have reduced temperature-related zero shift problems, but they are not as sensitive and cannot detect low ranges of pressure.
- Larger capillary tubing provides a faster response, but the increased volume results in increased zero shift due to ambient temperature changes. Smaller diameter tubing has reduced volume and tends to cause less zero shift, but the smaller cross-sectional area

increases the lag time considerably. This can be a big problem if the seal fluid has a high viscosity.

- Vacuum conditions in the process can ruin a seal, unless special hydraulic seal fluids are used. (Vacuum lowers the boiling point of the fluid and if the hydraulic fluid boils, the resulting vapor usually ruins the seal.) Some hydraulic fluids are designed to handle vacuum, but they tend to be viscous and may create other problems (see below).
- Choosing the proper seal fluid can be difficult. Trade-offs abound. Here is a quick list of things to consider:
  - > Some processes prohibit certain fluids (such as silicone, etc.) from being used because any leakage into the process would have undesirable consequences. Check with the plant to make sure this is not a concern.
  - > Low viscosity fluids provide much faster response and are usually suitable for lower temperatures, but they usually cannot handle vacuum or high temperature conditions.
  - > High viscosity fluids can handle higher temperature and vacuum, but they tend to have much slower response, and this response can get dramatically worse during low ambient temperature conditions.
- Be careful when trying to measure a low differential pressure between seals that are vertically far apart. (A common scenario is trying to measure the differential pressure across a distillation column.) In this scenario, the weight of the capillary fluid in the legs shifts the zero dramatically. Most transmitters will only allow a zero shift of four to five times the maximum range. If you are trying to measure 0-25″ wc across two taps that are 100′ apart vertically, the required zero shift will be approximately (100′ × 12″ × SG of fluid), which will be well beyond the zero shift allowed for most transmitters. A higher range transmitter can be used, but sensitivity will be lost.

Watch Out: Be extremely careful to select knowledgeable technicians to install capillary seals. Many a pipefitter has pulled them out of the box and bolted them up without the proper gaskets and spacers. If this happens, the seals will almost certainly be ruined.

Exceptions: If your process could encounter high vacuums at high temperatures, evaluate your options carefully. There may not be a fluid available that will suit your application.

Insight: Never use a single seal, pad type tank level DP transmitter on a tank whose temperature varies. This type has a single 3″ or 4″ seal on the high side and is vented to atmosphere on the other side. Because this arrangement has a large seal on only one side, the unit will be prone to significant zero shifts due to process temperature changes. If the process temperature fluctuates, the level reading will fluctuate as well.

Rule of Thumb: When faced with specifying this type of meter, you would be wise to seek out help from an expert until you fully understand all of the options and design trade-offs. These meters are NOT cheap and the specifying engineer can ill afford a couple of iterations to get it right.

Tip #15: You Cannot Control What You Do Not Understand

Hunter Vegas (System Design)

This tip seems obvious, yet some of the biggest project failures I have witnessed were a result of the team not understanding the process, and programming what they thought was correct. The folly of this approach was usually not discovered until late in the project (possibly during start-up). At that point, recovery of the schedule was impossible, and the effort required to correct the problem resulted in massive budget overruns.

Concept: Designing and/or improving a control system is practically impossible if the team does not understand the process. The first step of ANY control project should be a study of the P&IDs and a conversation with the process gurus of each affected area.

Details: Before beginning any major project, pull out the P&IDs, batch sheets (or other process documentation), and any operator instructions that can be found and read through them. Then take the time to talk through the process with the plant engineers and operations staff. Be sure to inquire about the normal process flow and any non-routine cleanout, emergency, or abnormal situations that may occur. If the process documentation is sparse or not updated, take the time to interview plant personnel and document the process and the functionality of the current control system. This thorough understanding of the process is absolutely critical to project success. Armed with this knowledge you can not only design the system correctly, but offer ideas for improvements that the plant might not have considered.

The project team's understanding is as crucial as the project leader's. If you are the project leader, distribute copies of the P&IDs to the entire team, and walk them through the process. Many project teams find it useful to map out the process on a large wipe board in the project team area during the discussions. This "process roadmap" can be left up for the entire project so team members can easily refer to it and discuss control options and ideas during the system design.

Watch Out: Always ask about "abnormal" scenarios. Clean outs, shutdown/start-up sequences, and other non-routine events may require more programming effort than the original project itself. They can also be the most difficult to program due to various interactions and undocumented operations.

Exceptions: There really are none. Even if the job is a control system retrofit with no significant software changes, take the time to learn the process.

Insight: In many systems, you will find programming errors that have been there for years. When asked the operators will often say, "Yeah, it has always done that, and we always wondered why." If the team encounters something that just looks wrong, ASK! Either the team does not understand the process nearly as well as they think they do or the original program was in error. Either way you are well ahead by raising the question.

Rule of Thumb: Process understanding is a crucial first step in any automation project. A failure to achieve that understanding can place the entire project in jeopardy.

"Thorough understanding of the process is absolutely critical to project success."

Tip #16: Use Outlines and/or Flow Charts First

Hunter Vegas (System Design)

As a student in high school, I struggled when writing papers. I had a lot to say, but the resulting jumble of thoughts and ideas was poorly presented and confusing to the reader. A ninth grade teacher hammered me for this, and ultimately taught me that no article or paper could be successful if it lacked a logical organization. He preached the concept of first creating an outline to assemble the main concepts in a meaningful and logical way, and THEN fleshing it out the paragraphs. This is a lesson I never forgot, and as I got into automation, I realized this same approach is just as critical to any kind of software development.

Concept: This applies to any major undertaking (such as writing a paper, making a presentation, or designing complicated batch code). Take the time to lay out a design and get the underlying structure right before you begin. You will save yourself hours of wasted effort, and your work product will be much more cohesive.

Details: When faced with a major creative endeavor, most people want to jump in and get started so they can feel that they are making progress. They sit down and immediately start writing/painting/banging out lines of software code. This method can eventually work, but the path to success will be a circuitous one full of wrong turns and rework.

Save the team a lot of wasted effort, and take the time to work out a solid framework before starting on the details. This concept applies to any major project but is especially true for software development. If you first outline the major components and think through how the parts will interact, the resulting code will be much simpler and testing and start-up will go infinitely smoother. Resist the urge to just sit down and bang out code. Many a project team has failed to do this and has blown the entire labor budget trying to patch and cobble something together only to ultimately step back, toss all the work done to date, and effectively start all over.

Once the design is complete, document it in a manner that is clear and unambiguous to the project team. The form of the documentation will vary depending upon the project size and type, but regardless, the documentation should be easily read and understood and easily updated as the project progresses to completion. This same documentation can be used for testing and checkout purposes at the end of the project and provided to the customer for future reference.

Watch-Outs: Due to tight schedules, many project managers encourage "concurrent engineering," in which multiple teams are working simultaneously rather than sequentially. Fight the urge to let the programmers "get started" while another group is working out the software design details. If the programmers are allowed to begin in advance with no direction, much of their work product will ultimately have to be reworked or simply abandoned.

When multiple people or vendors are part of project, spend extra time defining the boundaries where these groups interface. This is a common failure area of large projects.

Exceptions: Concurrent engineering is possible if it is done correctly. One portion of the design can be completed and released for detailed software development while the other areas are being designed and outlined. Keeping everything straight can be difficult, but it can work if the team communicates well.

Insight: This same advice is invaluable for writing a paper or creating a presentation. A simple outline will help focus the paper and make for a much more understandable document/talk. Start with a high level outline to list the concepts and the order of presentation, then add details to each section. Once a detailed outline has been created, converting it to a paper or talk is relatively easy because each line in the outline becomes a sentence or two in the final product.

Rule of Thumb: Always take the time to map out a cohesive structure before starting any creative endeavor. A sound outline or flowchart is a crucial first step for any paper, presentation, or software development project.

> **"Save the team a lot of wasted effort, and take the time to work out a solid framework before starting on the details."**

Tip #17: You Can Never Have Too Much Messaging

Hunter Vegas (System Design)

This is one of those tips that you cannot appreciate until you have worked on a control system that did NOT have adequate messaging programmed in it. Operators want to know what is going on, and system messaging is usually their only clue. Imagine running a complex or dangerous process and not having any feedback to tell you what is happening or if anything is happening at all! Or perhaps just as bad, imagine having the process sequence stop with such useless messages as "Sequence on Hold."

Concept: Nothing frustrates an operator more than working on a control system that provides vague, generic messages, or worse, provides no clue that anything is happening at all. Tell the operator what is going on! If a phase goes to hold, the system should tell the operator why. If the system is on hold for 10 minutes, the operator should see a countdown. The extra effort to do this is minimal, and the positive impact on operations personnel is immeasurable.

Details: Creating detailed messaging is easy once the phase templates have been configured to include it. The messaging can be displayed in a two line message bar at the bottom of the screen that is used for active phase messaging and for operator questions and responses. Two lines are usually necessary because many batch processes have multiple operations occurring simultaneously and providing two message areas avoids overlap.

Here is a sample of some of the messaging an operator should see:

- During timed holds, the system should provide a countdown: "10 minute hold, 6:33 minutes remaining."
- During material charges the message should indicate the amount and type of material charged and the amount remaining. For instance, the message might read: "Charging 200 gal caustic, 52 gal remaining."
- If a phase goes to hold, the system should always indicate WHY it went to hold in a detailed manner. For instance, the message might read: "Rx 300 to storage transfer held due to high level in product TK 301." Resist the urge to write generic messages such as "Phase on hold due to valve misalignment." Indicate WHICH valve is the problem.
- During equipment setup, the system should indicate what actions are being performed. For instance, the starting of an agitator might generate these messages: "Starting Agitator," "Ramping speed to 75%," "Agitator at Speed" etc. Generic messages such as "Setting Up Equipment" do not give much indication of what is happening.

Take the time to do messaging right. The improved operability and increased information to the operator will reduce downtime and allow the operators to quickly identify problems and resolve them without outside maintenance and/or engineering help.

Watch-Outs: If a programmer is not a good speller, do not have him or her doing the messaging! A couple of misspelled words will dramatically lower the perceived quality of the control system in the operator's eyes. (See Appearance Matters tip #30.)

Exceptions: There aren't any. Even a simple sequence can be programmed with a short message bar that is populated as logic is executed.

Insight: A trick to making messaging easy is to create one (or two) operator message variables that appear in the message lines on the bottom of the page. The various phases write to these variables so that the operator gets used to seeing phase related messaging in the same place. These message bars can be placed on several graphics as appropriate.

Rule of Thumb: Messaging can make or break a control system. If the information is detailed and useful, the operator will be able to run the system much more independently and resolve problems without any outside help. If the messages are generic or non-existent, Operations will be calling Engineering all night long looking for help to identify problems and get the process running again.

"Take the time to do messaging right."

Tip #18: Cheap Instrumentation Is Exactly That

Hunter Vegas (System Design)

The expression "You get what you paid for" definitely applies to instrumentation. Automation is ***expensive,*** but the payback is there when it is executed correctly. Unfortunately, many engineers try to save a couple of bucks and go with the untested "bargain brand" and often pay a wicked price for it. Some instruments are cheap for a reason.

Concept: Short term, penny-pinching instrumentation purchasing decisions may cost a plant thousands (or even millions) of dollars over the life of a control system. Spend the extra money and buy a brand that the plant knows and trusts.

Details: Automation engineers are under a lot of pressure to keep the price down, or they are asked to do a project for significantly less money than was estimated. (How many times has an engineer been asked, "I know the project was estimated at $yyyy, but what can you do for $xxxx?") In such cases, resist the urge to go with cheaper instrumentation and instead cut the overall scope to make up the budget difference. You can always find an alternate brand of instrument that is being offered at a significant discount and is marketed as being "just as good as Brand X." Unfortunately, the ultimate price that the plant often pays fighting through start-up, struggling to get technical support, and ultimately replacing the unit will be significantly higher than what would have been paid if the higher quality instrument had been purchased at the start.

Please note that this tip is NOT suggesting that cheaper equipment does not deserve a chance. Sometimes a vendor discovers a better technology or a cheaper method of manufacturing that DOES produce an instrument that is lower in cost and is just as good or better. In such a case, purchase a single unit and trial it in a less critical area of the plant. Buying 200 of the untested units for a large project is just asking for trouble.

Watch-Outs: Avoid sole sourcing instrumentation if possible. Even more importantly, avoid having a vendor that sells instrumentation and also provides the engineering to specify the instrumentation. The opportunity for a conflict of interest will simply be too great.

Exceptions: Some instrument brands stand so far above the others that it simply is not worth investigating another vendor. In such a case, bidding alternates is probably a wasted effort unless the alternate offerings improve dramatically or the sole source vendor starts raising their pricing significantly.

Insight: Buying high quality instrumentation does NOT mean that the buyer has to get fleeced and pay list price. Always have at least two major vendors qualified for each type and brand of instrumentation. The competition keeps both of them honest, and a plant can often achieve some good discounting if they are a consistent customer.

Rule of Thumb: Do not skimp on instrumentation. Buy good quality equipment and it will last for years. To keep the price down, always have two qualified vendors.

> **Short term, penny-pinching instrumentation purchasing decisions may cost a plant thousands (or even millions) of dollars over the life of a control system.**

Tip #19: Always Run Spare Wires and Plan for Expansion

Hunter Vegas (System Design)

Over my career I cannot think of a single time when I regretted running spare cables or oversizing field junction boxes. However, I can think of too many instances where I ran out of room and/or capacity much sooner than I had expected, and wished I had run MORE spares.

Concept: The incremental purchase cost of a 36 pair cable over a 24 pair cable is practically negligible when compared with the labor cost of running either cable. Install spare wire capacity whenever possible, or at least plan for future expansion when sizing junction boxes, cable trays, and control system I/O.

Details: During system design, always run spare cable, oversize the field junction boxes, and buy extra I/O cards. Ideally, the cost of this spare capacity can be worked into the project cost during the estimating phase so the money will be available when the project is approved. Even if the spare capacity was not included in the original budget, the additional cost of adding the extra capacity is usually low enough that it can be incorporated with no significant impact on the budget. If the project budget is so tight that larger cables or spare capacity cannot be installed, at least PLAN for future additions. Install slightly bigger field junction boxes and cable trays so that more cables can be added later. Oversize conduits, and leave a draw string in them so wires can be pulled in on future projects. Leave blank spaces in the I/O cabinets so I/O racks can be added later, or at least leave space in the room so I/O cabinets can be installed in a future project.

This concept is particularly true when running fiber optic cable. The labor to run a 6 fiber, 12 fiber, or 24 fiber cable is essentially the same and the material cost difference is low. Running a cable with less than 12 fibers is pointless, and if expansion is even slightly possible, run a 24 fiber cable.

Watch-Outs: Always ask about future expansion plans during the design phase of an automation project. By knowing how the ultimate system might appear, you can make minor design changes that will make future growth much less costly and difficult. If the system will double in size, you can lay out the I/O cabinets, power supplies, etc. in such a way that they can be easily upgraded in the future without burdening the current project significantly.

Exceptions: Occasionally an automation project involves a machine or a process that is so mature that future expansion is unlikely. This is not a common occurrence, but if this situation applies, clearly the cost of adding spare capacity would not be justified.

Insight: Project managers hate adding spare capacity after the project has been approved because they consider it "scope creep" and not part of the project. Plants adore spare capacity because it allows the execution of process improvement projects at a much reduced cost. In the long run, the company certainly saves money, but the project manager is not nearly as concerned with the long run as the execution of his or her project. If the spare capacity is included in the original project estimates, then all parties win. The project manager is happy because the project scope has not changed, and the plant and company are happy because they can expand in the future at a much reduced cost.

Rule of Thumb: A common rule of thumb for most systems is to add at least 25% spare capacity to the original design. Depending upon the future plans of the plant, a higher percentage might be justified. Adding less than that figure is usually not advisable unless the plant and/or process are mature and unlikely to change.

> **"Plants adore spare capacity because it allows the execution of process improvement projects at a much reduced cost."**

Tip #20: Spend the Money to Individually Fuse I/O with Indicating Fuse Blocks

Hunter Vegas (System Design)

I recently encountered a control cabinet on a client's site that had 50 valve limit switches wired to a single breaker. If any of the 50 shorted, the entire group tripped off line. Between this installation and several others like it, the technicians often spent hours and even days trying to track down a single field wiring problem while the process equipment sat idle.

Concept: The small amount of extra money required to individually fuse I/O with indicating fuse blocks will be quickly recovered through dramatically improved troubleshooting. Even current limited I/O can benefit from individual fuses and/or disconnects.

Details: Troubleshooting a system that does not have individually fused I/O can be one of the most difficult and time-consuming activities for a technician. A single fault in the field can take out all the points on a card and might take out the entire cabinet. In such a case, the technician usually starts lifting wires and continually resetting the breaker until he or she finally isolates the problem. This problem can be avoided if the cabinet design incorporates individually fused I/O and indicating fuses. If the cabinet incorporates these features, the problem is immediately obvious the moment the door is opened. (The blown fuse light is difficult to miss.)

Unfused I/O is common in third-party skid packages where the vendor is trying to reduce costs wherever they can. Eliminate this possibility by specifying that all I/O must be individually fused with indicating fuses in the bid package.

Watch-Outs: In an attempt to add fuses, avoid the temptation to use two-, three-, or four-high terminal blocks to save room. These blocks look wonderful on paper but are AWFUL when they are installed in the field. The technicians cannot even SEE the lower terminals, much less get their probes on them for voltage readings, and the cabling is an absolute nightmare.

Beware of I/O cards that purportedly include individual fuses. Almost none of these cards use indicating fuses and some of them require the entire card to be removed in order to replace one fuse!

Exceptions: Some I/O cards utilize current limiting circuits that can sustain a field short without damaging the card. In such a case, fuses are unnecessary, but consider installing terminal blocks with a built in disconnecting plug. Such a disconnect provides an easy means for the technicians to take series current measurements or connect their handheld communicators.

Insight: The increased cost of using individual fuses will be quickly recovered by the reduced time to troubleshoot and resolve field wiring problems. One or two instances of bringing production back on line within minutes rather than hours (or days) will easily pay for the initial installation.

Rule of Thumb: Individual, indicating fuses and single-high terminals take up more room in the cabinet, but the benefits far outweigh the costs. Build this into your standard cabinet designs and be sure to specify it in your third-party skid package specifications.

> **"Troubleshooting a system that does not have individually fused I/O can be one of the most difficult and time-consuming activities for a technician."**

Tip #21: Too Many Alarms Can Be Worse Than None at All

Hunter Vegas (System Design)

At one time, I worked in a large continuous process plant that had alarms coming in constantly. The operators could hit the "Silence" button in their sleep. We had a case where a process flow was accidently diverted to the wrong tank, and it eventually filled and overflowed the tank. Even though the tank had redundant level transmitters and we had one of the more alert panelboard operators on shift, the rising level was not noticed until the tank overflowed and was noticed by a field operator. The panelboard operator had silenced two high alarms, two hi-hi alarms, two "over" alarms, and two range alarms over the course of two hours but had failed to recognize that there was a problem.

Concept: Enable alarms on instruments that matter and on process nonconformances that the operator can do something about. Having alarms for the sake of having alarms only ensures that ALL alarms will be ignored—even the ones that matter.

Details: Alarm management has become all the rage lately, and with good reason. The proliferation of instrumentation busses has provided access to a plethora of information, and because everything is now typically alarmed, the operators are being buried under a barrage of alarms. When faced with a constant stream of annunciation, most operators quickly become numb and increasingly just hit "Silence." Critical alarms are lost in the noise and are routinely missed.

Ironically, there was an advantage to the relay annunciator panels built into the old control room panelboards. There were only so many points available, so only the critical alarms made the list. With the advent of computers, EVERYTHING can be alarmed, and unfortunately that is exactly what happens.

An engineer has several ways to address this problem, and many books have been written on the subject. Addressing this expansive topic in a few pages is not possible, but here is a brief list of suggestions that can help reduce the problem of too many alarms.

- Enable alarms on instruments that matter.
- If an operator cannot do something to resolve the situation, there is no point in alarming it.
- Program "smart" alarms. Automatically disable alarms on out-of-service equipment. Add conditional logic that generates a common alarm when a piece of equipment trips rather than generating 10 or 15 alarms that essentially indicate the same condition. (For instance, if a boiler trips it makes little sense to alarm the trip, low gas flow, low gas pressure, low air flow, etc.) One piece of information that IS useful, however, is "first out" trip information. Many operators use the alarm list to determine what tripped the equipment. If the first out information can be indicated on a graphic, the operators do not need to see the individual alarms.
- Segregate the alarms and deliver the information to the appropriate audience. The operators do not need to see most calibration and/or maintenance alarms, but the

maintenance department does. Generate an alarm report to Maintenance, but just indicate a possible problem to the operator so he or she can be aware of it.

- Change from alarms to indicators. If a process is running out of spec but not in a critical range, then it may make more sense to indicate this condition as a color change on the graphic instead of than firing an alarm that must be acknowledged.
- Monitor alarms and routinely eliminate "bad actors." In most cases, a large percentage of alarms is created by a handful of points. An occasional review of the most active alarms will allow the plant to identify these points and modify the programming to reduce their frequency or address their cause. Doing this can dramatically reduce the total alarm count without requiring much effort.

Watch-Outs: Many control systems default to having all the alarms enabled. On a new system, it may make more sense to enable none of the alarms initially and add them back as necessary.

Exceptions: Some plants do not allow the operators to suppress alarms because they are concerned that critical alarms will be turned off and never restored. One solution to this problem is to allow operators the ability to suppress alarms, but program the alarms to automatically restore after some appropriate period of time. In this way, a broken instrument can be silenced for a shift while repairs are made.

Insight: One plant only enabled setpoint alarms when a controller was in automatic. (Such alarms annunciate when the process variable is beyond the allowable range around the current setpoint.) High and low alarms were not enabled unless the controller was in manual. This method provided increased alarming when a loop was in manual but did not generate alarms on a point in automatic unless it deviated too far from setpoint.

Rule of Thumb: Alarm management is a never ending effort. Routinely review the plant's alarm list, and try to eliminate or address points that appear too often. When configuring new systems, include some means of smart alarm management into the design.

> **"With the advent of computers, EVERYTHING can be alarmed, and unfortunately that is exactly what happens."**

Tip #22: Details Matter

Hunter Vegas (System Design)

I was working on a large automation retrofit project of a chemical plant that had numerous thermocouples scattered throughout the structures. A vendor suggested a particular thermocouple card that I had never used, and I was about to proceed with that card when I decided to quickly scan the specs on the card. There was an odd footnote about channel-to-channel isolation that caught my eye, and as I dug deeper into the details, I realized that channels 1 through 4 and channels 5 through 8 shared a common ground on the card. Now, this plant was older and had a mix of grounded and ungrounded thermocouples. If I happened to get thermocouples from two different columns on the same group of channels, the resulting ground loops would have sent the readings all over the place. Luckily, I noticed this and was able to pursue a different path early in the design process.

Details matter—take the time to chase them.

Concept: If you are not a detail person, either pursue a different career or LEARN to be a detail person. The field of automation demands extreme attention to detail. Wire size, materials of construction, flange size, flange rating, min/max flows, temperature, pressure, etc.—the list goes on and on. Failure to evaluate even one item can have serious or even catastrophic consequences. Check and recheck everything— "good enough" can shut down a plant or even get people killed.

Details: Engineering is by definition a detail oriented profession, but the field of automation requires almost fanatical attention to detail. Everything matters, which is why instrument spec sheets have so many lines on them. Engineers who just copy/paste spec sheets from a similar transmitter or use a control panel design without understanding it will not last long in this field. Automation is particularly challenging because the engineering skill set is so diverse. In the normal course of a job, automation engineers might find themselves doing mechanical design, electrical design, and chemical process design, and each of those has a long list of details to consider. Obviously, a comprehensive list of the necessary details in the field of automation would take multiple books to cover and cannot be provided here. However, here is a list of examples that will help you learn to chase the details:

- You should never copy or modify a design that you do not understand. A panel hardware design may involve a hundred design decisions—wires and I/O racks are sized for current and future needs, space is left for thermal dissipation, wireway is sized for current and future field wiring, locations for conduit penetrations are allocated, the cabinet is sized to fit in the room and even sized to fit through a particular doorway. If you copy that design and modify it without understanding what went into the original design decisions, you may quickly find yourself in trouble. The panel equipment may start overheating because a larger power supply was specified, or the main breaker may blow on start-up because the inrush currents are higher, etc.
- Read the specifications and understand them. Failing to consider the pressure/temperature limitations on instrumentation can have catastrophic consequences. Failing to look at the electrical details of I/O cards (grounding, leakage currents, voltage limi-

tations, etc.) can turn a start-up into a nightmare. Looking at spec sheets might be the very definition of boredom, but FAILING to look at spec sheets may generate much more excitement than an engineer wants to encounter.

- Learn to color. Young engineers are notorious for working fast but missing details. One of the best ways to address this problem is to pick up a highlighter and learn how to color the lines as you work. If you are checking drawings, then highlight each section as it is completed. If you are doing an instrument takeoff, then color each instrument on the P&ID as it is added to the list. In short, find a document that covers everything that needs to be done, and color it as each item is completed. Checklists (such as those found at the end of this book) can also be an invaluable way to make sure everything has been considered and evaluated.
- Trust, but verify. After an engineering team leader has worked with a team for any length of time, he or she gets to know the strengths and weaknesses of each team member and knows where and to what degree back checking is required. New engineers should probably be 100% checked initially but can be backed off to spot checking once their skills are proven. However, even work done by seasoned engineers should be reviewed to some extent because everyone makes mistakes.

Watch-Outs: Always cross check vendor sizing calculations for instrumentation. Vendors often plug information into sizing software and generate impressive specifications and calculations, but they do not know the process, and they make errors. It is always worth running a rough cross check on their sizing and then reading through the entire specification to make sure the materials of construction are as required for the application.

Exceptions: None.

Insight: Many large engineering firms send a spec sheet with a smattering of process information to the vendors and let them generate the instrument specifications. This practice invites disaster. The vendor cannot possibly know the process details or the abnormal conditions that the instrument might encounter.

Rule of Thumb: Successful automation engineers HAVE to be detail oriented. If this is a not a natural tendency, then learn to color, cross check, or do whatever is required to be detail oriented.

Tip #23: Anticipate Murphy; He Is Alive and Well

Hunter Vegas (System Design)

We all know Murphy's famous law, "If anything can go wrong, it will." I have to believe that Murphy was an automation engineer because I have encountered his law in action on every project I have ever worked on.

I have sat in HAZOPs where the group wanted to discount a scenario because it involved two simultaneous failures. I have also worked in a chemical plant that encountered FIVE simultaneous failures, blew up a vent line, and narrowly missed injuring an operator. Equipment breaks, and people make mistakes. Anticipate it, and design for it.

Concept: Simple systems work reliably. Complicated systems find new and interesting ways to fail. Whenever possible go for the simplest, most robust solution. As an automation engineer, the KISS concept (Keep It Simple Stupid) should be your mantra.

Details: Automation engineers love to create gloriously complex solutions. With so many computers and gadgets available, it is hard NOT to want to incorporate the latest and greatest into a design. However, the true purpose of automation is to control the process. Sometimes it takes a multivariable predictive control model to do that, but many times it can be done with a float switch and a solenoid. Try not to complicate a solution any more than necessary. When you are designing an emergency system to dump a quench chemical into a reactor, consider using gravity rather than special pumps and other equipment. Gravity always works (at least on planet Earth), while pumps and/or electricity can fail—especially under emergency conditions.

Anticipating every failure is difficult, but you must make every effort. What happens if the operator presses the wrong button? What happens if no button is pressed at all? If power is lost, might the instrument air and cooling water systems fail as well? What about steam and nitrogen? What are the ramifications of these multiple failures?

When you are designing a control panel, consider using dual 24VDC power supplies. Feed one with a UPS circuit and the other with a non-UPS circuit. Despite what their name might imply, an Uninterruptable Power Supply becomes an Interruptible Power Supply more often than not. Having dual feeds can allow a control panel to continue operating despite the failure.

Software design is particularly tricky because there are so many paths that the logic can traverse. Operators are forever using the equipment in ways that were never intended and if the software is not designed to handle it, the program can hang in unexpected places. During testing try hitting the wrong buttons and try to force the program to step through the sequence in a different way to see what happens. While this will drive the programmers crazy, the resulting system will be much more robust as a result. Finding and resolving problems in testing is always better than discovering them on start-up!

Watch-Outs: Never allow the final software quality control testing to be implemented by the same person who programmed it. A different person is much more likely to hit the sequences in a different way or throw the system a curve that the programmer had not anticipated.

Avoid the temptation to use exotic controls and programming to patch a poorly designed process. You can program around poor mechanical designs, but the project will be more stable if the fundamental problems are resolved.

Exceptions: Sometimes a HAZOP group can lump a series of totally improbable scenarios together and reach outlandish conclusions. However, there ARE certain scenarios that can create a cascade affect. (A loss of power might trip the steam system and take out the cooling water supplies as well.)

Insight: Safety interlock calculations include a testing interval and incorporate the failure modes into the calculations for a very good reason. Untested interlocks have caused hundreds (and probably thousands) of accidents when they failed to perform their function. Be particularly wary of interlocks that involve multiple instruments and/or devices to sense a failure. The probability of failure on demand will be very high.

Rule of Thumb: If you are given an option, always choose the simpler solution. When you are designing a system, do not consider operator error and equipment failure to be isolated and unlikely events. They will occur ... and usually at the worst time possible.

> **"Whenever possible go for the simplest, most robust solution."**

Tip #24: When Performing a DCS Retrofit, Take Time to Study the I/O Cards

Hunter Vegas (System Design)

This is a tip that I simply cannot stress enough. If a team does not understand the intricate electrical details of the old control system and the new, really REALLY bad things can happen.

Concept: Performing a control system retrofit without completely understanding the electrical characteristics of the old and replacement system I/O cards is a fast track path to disaster. Subtle differences can create start-up nightmares.

Details: This tip is really an extension of Tip #22 ("Details Matter"). When an engineering team is faced with the task of replacing an existing control system with a new one, the first step should be an exhaustive study of the I/O card designs in the existing system and in the proposed replacement. Any differences should be investigated in detail. Here is a partial list of the items that should be checked:

Analog Cards:

- Are the points grounded, or isolated from ground? (Grounded analog input cards can create ground loops, and if the point is tied to ground, wiring analog signals in series with other systems may not be possible.)
- Do the analog cards source voltage, or not? (Some systems allow selection of two wire or four wire connections on a point by point basis. Does the replacement system have similar options?)
- Are the points on a given card all the same type? (One older system allowed the user to set each channel on a card to be 2 wire analog inputs, 4 wire analog inputs, analog outputs, pulse inputs, and even 24V digital inputs and outputs – few new systems can match that functionality.)
- What is the impedance of the analog input cards? (Older systems may have had low impedance –50 ohms or less.) New systems tend to have 250 ohms to allow HART communications. If field indicators are present, the additional impedance can pose voltage drop problems.
- How are the shields landed and grounded? (Different cards handle the shields differently.)
- Do the analog cards measure the current LEAVING the system, or returning to it? (If the old system measured the current going to the field, then transmitters will continue to function even if their return wiring is shorted to ground. Such a ground will keep the signal from reaching the DCS if it measures the return current.)

Digital Cards:

- The voltage ratings must match. Obviously, using 120VAC cards for 24VDC signals would not be a good choice.
- What is the current rating of the outputs? Are the points fused? (Be sure to check the individual channel rating as well as the rating of groups of channels on a card. Some

cards have a high per channel rating but have a comparatively low rating for the collective current of all of the points.)

- Does the digital input card source or sink voltage? Some cards provide voltage that loops through the field contact and returns to each channel. Others provide voltage on a channel by channel basis and look for the field to ground it out. Still others provide no voltage and expect it to be provided externally. Some cards can take voltage from isolated sources, while others require all channels (or channel groups) to use a common voltage. The new system must be compatible.
- Does the digital output card source or sink voltage? A relay card can generally switch externally powered 24VDC and 120VAC signals on a channel by channel basis. Other DO cards provide voltage on each channel that may be fed from external sources or fed from the card. Still other cards ground out each channel, providing a current path for externally fed power. As with the digital input cards, the cards come in many different flavors that may or may not be compatible.
- Beware of leakage currents. Some digital input cards will turn on when an input receives even a tiny amount of current (<1mA). If a field device leaks current (as many proximity switches do) or if the signal is fed from another system digital output card that leaks 1 or 2 mA of current, the digital input will turn on and never turn off.

Intrinsically Safe Systems:
If the automation retrofit project involves existing IS barriers then the opportunity for error is huge and the consequences especially severe. A book could be written on the subject, but as a start here is a quick list of things to consider:

- Voltage drop – IS system voltage drops are high in general, but if the new system has a higher impedance, then the existing barriers will probably not work.
- Grounding – Some barriers are grounded internally. If the system measures returned analog signals, then they may not reach the system if the barrier is grounded.
- Supply voltage – Some barriers use an elevated supply voltage to provide more voltage to the field. Is this compatible with the new system?

If the project plan calls for the reuse of existing barriers, then the team would be wise to obtain a sample of each barrier and test it with the new control system I/O cards. Barriers may not function when they are wired to the new system.

Watch-Outs: Many vendors provide interface hardware or special connectors that supposedly allow the user to "instantly" connect the existing I/O into the new system. Evaluate these offerings carefully. In many cases, the opportunities for hardware incompatibility mentioned above still exist and need to be evaluated. (Differences in grounding, voltage drops, impedances, etc. can still wreak havoc.)

Exceptions: None.

Insight: Start with detailed schematics of the old system and the new. Investigate current limitations, leakage currents, impedances, and particularly grounding.

Rule of Thumb: The electrical analysis of the systems should NOT be assigned to a junior engineer. If just one detail is missed, whole banks of I/O may fail to function on start-up.

Tip #25: Make Sure the Coffee Is Strong and the Pot Brews Quickly on Start-up

Hunter Vegas (Project Management/Execution)

This may seem like a crazy tip, but there is a lot more truth in this statement than you might realize. It's directed to management, but a young engineer may have reason and opportunity to encourage management to do this.

Concept: Simple things like strong coffee and a fast brewing pot seem minor, but during start-up or production outages they can impact productivity and morale a lot more than you might realize. Bringing in a couple of pizzas or a few half gallons of ice cream costs next to nothing, but it can make a world of difference in the attitudes of the people who are working 24 hour coverage to get a plant started up or back on line.

Details: A start-up or production outage is a hectic, chaotic time. People may be working seven 12s for days and weeks at a time, patience is short, tempers flare, and the pressure mounts as everyone struggles to stay on schedule and resolve the myriad of unexpected problems that invariably crop up. With so much going on, anything management can do to ease the situation and improve morale is a worthwhile thing. Bringing in donuts in the morning or pizzas or subs at lunch is definitely appreciated by the crew. Some companies have "ice cream socials" during the shift change meeting. Do SOMETHING to show that the company appreciates the extra effort.

Watch-Outs: Do not forget the night crew, the off shifts, and even the contractors. Arranging food for these groups can be logistically more challenging, but the negative effect of NOT including them is great – not to mention being unfair to them.

Exceptions: The biggest risk to this idea is setting a precedent where people simply begin to take the food for granted. If possible, arrange the special meals irregularly enough that their arrival is still a surprise.

Insight: As crazy as it might seem, a fast brewing coffee pot and easy to use coffee pouches probably create a measurable improvement in the productivity of an engineering office as well. It would be an interesting study to find out how much time is wasted sitting around the coffee pot waiting for it to brew and talking about last night's ball game.

Rule of Thumb: Take the time (and spend the money) to show appreciation to the crews on a start-up or during a major production outage. These people are under a great deal of stress and deserve the recognition.

"Bringing in a couple of pizzas
or a few half gallons of ice cream
can make a world of difference..."

Tip #26: Document Your Code

Hunter Vegas (Project Management/Execution)

You will invent a whole new library of curse words the first time you have to come behind a previous engineer and work on a program that involves thousands of lines of complex, undocumented code.

Concept: Documenting software code does not take a great deal of effort, but most programmers are loath to do it. Despite how they may feel, insist that all code include at least a minimum of documentation.

Details: Everybody understands the code that THEY wrote, but eventually some poor soul has to tweak or troubleshoot that code years after it was written, and he or she may be in for a battle if the original software was never documented. Occasionally, Karma prevails, and the original programmer has to go back and rework his own code 10 years later. Even though he wrote it, the programmer will often find it impossible to remember his thought processes 10 years after the fact.

It does not take that much effort to create meaningful variable names and descriptions and drop in a comment or two that explains what is happening in the software. However, it can take days (or weeks) to try to understand a complex piece of code and figure how to modify it as necessary. Programmers, take the extra few minutes and explain what is going on. This is especially important if the code is involved or uses subtle "tricks" to accomplish a particular task.

When creating variable names and descriptions, try to reference the field tag number and its function. This makes the tag much easier to recognize. In addition, try to be as consistent as possible when assigning these descriptions. Working on a system that was programmed by multiple people can be extremely difficult if every section reads and is documented differently.

Watch-Outs: Many third party vendor packages incorporate a PLC with no documentation whatsoever. Insist on documented code in your original bid spec.

Exceptions: None.

Insight: Back in the dark ages, PLC memory was expensive, and programs were written to conserve that memory as much as possible. Programmers used software pointers, tables, and rack addressing to accomplish certain tasks and most failed to document anything. In many cases, the slightest modification to the program could shift the registers, and suddenly everything would be pointing to the wrong place. It was great job security for the guy who wrote it, but they were rarely asked to write anything else!

Rule of Thumb: Take a moment and sprinkle comments through the software during programming. Ten years from now you (or someone else) will be very glad that you did.

> “Occasionally, Karma prevails, and the original programmer has to go back and rework his own code 10 years later.”

Tip #27: Create, Thoroughly Test, THEN Replicate

Hunter Vegas (Project Management/Execution)

Over the course of many projects, my project team and I have learned this tip the hard way, and with every project we promise to test new or revised software harder and more thoroughly than the time before. Despite all of our efforts, we STILL find small bugs late in the game that require us to touch each and every module to correct. I do not know if we will EVER learn this lesson completely, but we have gotten much more vigilant in our early testing. Pain has been very instructive.

Concept: 100 copies of garbage is a LOT of garbage!

Details: Many young engineers are in a hurry to show progress so they create a template, decide it works, and promptly make 100 copies to show their boss what they have accomplished. Later a bug (or two or three) is noticed, and suddenly they have two or three hundred bugs to fix rather than just two or three.

Earlier tips highlighted the fact that the automation field demands fanatical attention to detail, and this concept is a clear example. When creating a piece of code that will serve as a template for others, beat on it mercilessly to ensure the code is bug free. Once that is accomplished, give it to someone else and let THEM beat on it some more. Test it in every way possible, and make sure all of the problems have been wrung out before releasing it for replication. Doing this takes more labor initially, but it will save hours, days, and even weeks of labor further down the road.

This process seems like a simple task, but actually accomplishing it on a routine basis is always difficult to do. When the project is starting up and lots of activities are going on, finding the time and discipline to thoroughly debug code prior to replication can be challenging, but the consequences of NOT doing it can be dramatic.

This same concept applies to higher levels of programming. When programming a system with five virtually identical reactor trains, create and thoroughly test the software for one train in its entirety before replicating that software to the other four trains. If possible, create the first reactor train software, test it as thoroughly as possible, and then give it to the client for still further testing and review. Once everyone is satisfied with reactor train #1, THEN replicate the software for the other trains.

Watch-Outs: Have someone other than the original programmer test the code. As mentioned previously, a second person is much more likely to spot errors the programmer has overlooked.

Exceptions: None … unless the project is a time and material job and maximizing billable hours is the primary goal. (Just kidding!)

Insight: Once a piece of code HAS been tested and proven, consider moving a copy of it to a system library where others can use it. Doing this can save hundreds of hours of development work and reduce project risk.

Rule of Thumb: If a piece of code is to be replicated many times, always have two different people independently test it whenever possible.

> **"100 copies of garbage is a LOT of garbage!"**

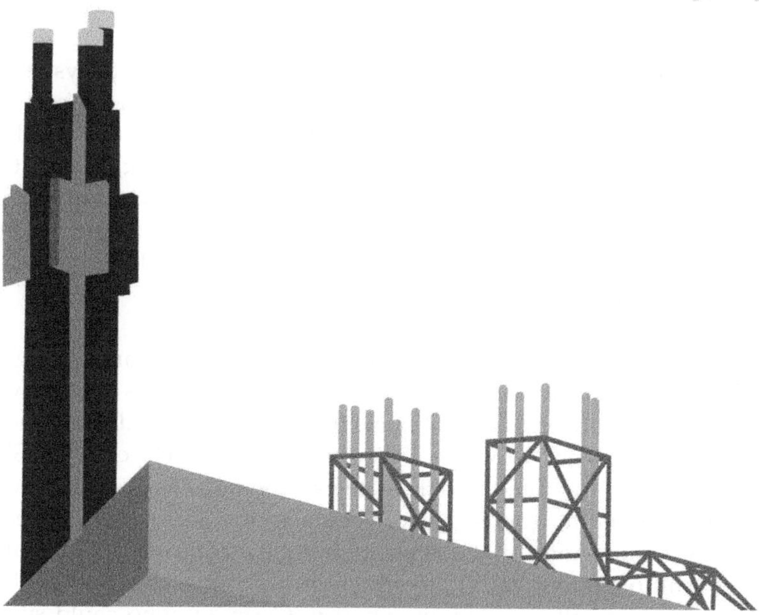

Tip #28: "Plug and Play" Devices Often Don't

Hunter Vegas (Project Management/Execution)

Over the course of many years, I have learned to be suspicious of anything that is marketed as "Plug and Play." I suppose I may be the only person with this problem, but with amazing regularity, I find plug and play devices that do not plug and play.

Concept: Control system communications can be a tricky thing. In some cases, the wires are hooked up, and everything begins communicating immediately. In other cases, NOTHING works, and it can take hours, days, or weeks to get the first piece of data through.

Details: It is thrilling to browse for a network printer, find it, and suddenly be able to print without any further problem. Unfortunately, more often than not it does not work that way. The computer cannot find the printer, or the printer driver cannot be found, or the driver is not compatible with the computer's software, etc. In short, it can take a great deal of time and effort to get "plug and play" devices to actually play.

With the advent of instrumentation networks, many devices and control networks are also marketed as "plug and play." Presumably, one can just hook up a device, and it will immediately start communicating. However, real world experience suggests that it is not always so straightforward. In my experience, either it works just as marketed and the system communicates immediately, or it will take hours, days, or even weeks to get online … especially if the technology is unfamiliar. Once communications are established, the networks generally work well, but getting to that point can be a trying experience.

If the project calls for the incorporation of a new network type into a control system, fully test it BEFORE the start-up begins. Gather a couple of devices (ideally one of each type), and set up a temporary network in advance to establish communications and get the network configured. Even if the project is just extending an existing network, be sure to pretest any new device types that are not already functioning. New devices usually require a different communication file or may need a special character string to get them to respond on the network. Even new versions of equipment that have been used for years can be troublesome if the hardware revision has changed.

Watch-Outs: On one project, the team added a new positioner to an existing Profibus PA network and did not test it in advance. Once the plant was shut down, the positioner was wired into the network, and problems immediately ensued. First, the positioner was a new model and was not listed as a choice on the system. The correct GSD was found, but then it had to be incorporated into the gateway configuration without deleting the existing setup and rebuilding it from scratch. Eventually, that was accomplished, and the device was finally communicating on the network, but the unit ignored the commands being sent to it. After multiple phone calls to several tech support sites across the United States and Canada, the team finally located an expert in Germany who could speak no English but did know the product. Emailing through a translator, the team finally realized that the positioner required the system to write a hexadecimal command to a certain register in the positioner so that it would accept the system command value as legitimate.

The device literature marketed the positioner as a "plug and play" device.

Exceptions: Some communications networks really do work right out of the box. However, a project team would be wise to plan for the opposite.

Insight: About a week before every major DCS start-up, consider performing a "network startup." During this time, power up as many of the systems as possible, and bring as many of the communication systems on line in advance of the plant cut-over. This early effort is critical because it gives the team a full week to resolve any fiber or major communication problems before the plant comes down. If communications are not established in advance, the whole checkout can be stalled while engineers frantically try to establish system communications.

Rule of Thumb: Find a way to test any communication network prior to installation. It may work perfectly without any issues at all. But when it fails to work, it can take a great deal of additional effort to get it functioning.

Tip #29: Tricks for Control Panels

Hunter Vegas (Project Management/Execution)

Project management is really just risk management, and a large source of risk to an automation project is the control panels. If these panels and the software can been fully tested before the team arrives on site, the only remaining unknown will be the field wiring. Here are some tricks to make sure your panels are ready to go.

Concept: A poorly designed or fabricated control panel can bring an automation project schedule to its knees during a start-up. Take the time to ensure that the panels are thoroughly tested and ready to go in advance.

Details: Most system integrators push the client to let them design and fabricate the panels at their shop so they know they can be thoroughly tested before arriving on site. A client is wise to agree to this arrangement, provided they are allowed to take part in the panel checkout and that the checkout occurs PRIOR to shipping the panels on site. This timing is important because if any problems are encountered during the checkout (wrong type or missing panel components, design problems, etc.), the client can get them resolved BEFORE the panels arrive on site. If the client does not get to review the panels in advance and problems are discovered after the panels are shipped, the client does not have much time to fix them and his negotiating position to deciding who will pay for the changes will be considerably weaker.

Here are some tips to consider when doing a panel test:

- If the panel is to be fed with two independent sources of power (such as UPS and non-UPS), connect the panel to two power sources that are out of phase. (If undocumented or unplanned links occur between one power feed and the other, it will quickly become obvious!)
- Tug (HARD) on every wire. This may seem like a colossal waste of time and can be a laborious task, but the odds are that you will find at least a couple of loose wires in the panel. If the wires are heavier gauge and jumpered, the copper tends to relax after the initial termination, and MANY wires will pull out. In addition, some electricians use an undersized screwdriver to do their terminations, and this becomes obvious when every other wire suddenly starts coming loose. (As a side note, most electricians get embarrassed when the engineer starts finding loose wires, and they will start torquing the wires hard from that point forward—a win-win for all!)
- First check the power supply wiring, then gradually power up the panel. Always perform a resistance check before energizing a breaker or inserting a fuse. Once the 120VAC power is fully energized, test and power up the 24VDC system. If the power feeds are jumpered, be sure to check that power is available along the whole strip. A single loose jumper can interrupt power to a whole strip of I/O.
- Energize and establish communications. If fiber transducers are used, put them in the network and use fiber jumpers to simulate the field fiber cable. Try to include every component in the system so that its functionality can be proven.
- Test each I/O point with real-world signals that actually draw some current. (A high resistance termination on a digital output will pass a voltage check but will fail under

load.) Use a 2W/4W analog simulator to test each analog loop. Use a small relay to test digital outputs. Feed voltage into every digital point. Do not rely on the indicator lights on the card. Some failures will allow the card I/O to work and the lights to function, but the information still fails to get transferred to the software.
- If the system has redundant power feeds (or 24VDC power supplies), test them individually and then test them with both sides energized. Trip each feed and make sure the system continues to function and the appropriate alarms sound.
- With as many digital and analog outputs turned on as possible, trip the main power feed and then restore it. This tests the current inrush of the system. A poorly designed panel may trip the main breaker under this condition.
- Once the panel is fully tested, REMOVE ALL FUSES (or at least those feeding the power supplies) BEFORE SHIPPING THE PANEL TO THE SITE! (See Watch-Outs below.)

Watch-Outs: Do not make the mistake of shipping a panel to the plant site with the fuses still in it. The electricians will hook up the power and will want to "smoke test" the system to make sure it works. If the power is hooked up incorrectly, they WILL smoke test the system and destroy several components. Alternately, they will power up the system and proceed to blow every single I/O fuse as they pull and connect wires in the field. As a minimum, pull the fuses that feed the main power supplies and I/O power. If power is put on the system, it cannot get far enough to do much damage.

Exceptions: In some retrofit projects, there are existing cabinets that must be gutted and have the internals refitted with the new system. In this case, prefab the panels as much as possible, and size them to fit inside the existing cabinets. This will still allow the team to test the panels thoroughly and the change over at start-up should proceed faster because the team is just "ripping and replacing" the internals rather than trying to install panel components during the shutdown. Be sure to measure the internal dimensions carefully, including the size of the doors. Many replacement panels have been fabricated to the wrong dimensions and would not fit in the cabinet. Still others have been fabricated to the correct dimensions, but the team failed to realize that the DOOR OPENING was smaller so they could not get the new panel through the door!

Insight: In a retrofit project, the cabinet design and the logistics of changing it out will have a huge impact on the length of shutdown required. While there is a great desire to re-use cabinet designs, realize that one size rarely fits all in system retrofits. In many cases, the cabinets and/or panels must be specially designed to accommodate the shutdown schedule dictated by the plant.

Rule of Thumb: Always take the time to thoroughly test control panels prior to shipment from the fabricator. A checkout should including pulling on each wire, performing "real world" tests of the I/O, and load testing the power supplies to make sure that everything is ready for installation. Then remove the major fuses and ship the units to the installation team.

Tip #30: Appearance Matters

Hunter Vegas (Project Management/Execution)

I was working on a large project early in my career and due to a last minute design change I had to quickly fabricate a panel with a dozen three way valves to allow Operations to manually control some equipment. I threw together a design and the contractor did a good job of procuring the panel and installing the valves. I then had a dozen labels made, indicating the operations of each valve, and gave them to a technician to install on the panel. Later, my boss passed by and commented on what an awful job I had done on the panel. I was flabbergasted and immediately went to the panel to see what was up. The labels looked like they had been put on by a four year old. Not a single label was parallel or centered or leveled. I ripped them all off, ordered a new set, and put them on myself. I learned a valuable lesson about the importance of a first impression and have never let anyone put labels on my panels since.

Concept: From the clothes you wear, to the labels on a panel, to the layout and fonts on a graphic, to the look of your signature – appearance matters, and the first impression you make will likely stick for a long time.

Details: This is a simple concept with a lot of applications, so it is easier to just list a few:

- When someone meets a person for the first time, they immediately make several snap judgments about that person that will remain in their subconscious for a long time. Was the person dressed neatly? Did he or she look them in the eye and have a firm handshake? Did the person talk clearly and with confidence, or did they mutter? If a person fails to make a good first impression, that person may spend weeks, months, or a lifetime trying to recover.
- When a client looks at the outside of a control panel, he can usually only see a couple of buttons, maybe a couple of lights and labels. If those labels are crooked or poorly made or the buttons are not aligned, the client will immediately judge the whole panel to be poorly designed or constructed even if the inside is an electrical work of art and functions flawlessly.
- When an operator is first introduced to a new control system, they will probably judge it within the first two or three minutes. If the graphics look clean and the buttons they push work as expected, that operator will consider it a "good" system. If the items are poorly aligned on the screen or the colors are wrong or the buttons fail to function or throw up an error, the operator will think the system "sucks."
- Never, EVER send a new graphic to a client until it has been thoroughly reviewed and tested. If possible, be present when he first sees it so that you can show him how things work and ease his mind.
- Despite a project's eventual performance, it is often judged by how cleanly it starts up. Good project management will eliminate most bugs before start-up. Be sure that you have resources to quickly correct any problem before it spoils the reputation of you and your project.
- It seems silly, but people will even judge other people based on their signature alone. John Hancock had a point when he signed the Declaration of Independence. How

many people can remember even one other signature on that document? A large, bold lettered signature exudes confidence. A small, diminutive signature suggests the opposite. Practice your signature until it looks impressive. Always fill the space provided, and even go outside the lines. Although it seems like an absurd thing, it really does make a difference.

Watch-Outs: At the beginning of a job a client will often pressure the integration team for some early graphics for review. The liaison engineer will often promise that only he will look at it. Do not transmit ANYTHING to the client until it has been thoroughly tested and it works as it should. This early work will often get shown to a couple of operators. If the work is not correct, it will make a bad first impression that will be difficult to overcome.

Exceptions: If the client is new and a set of graphical templates has not been established yet, send several different generic graphics with different template options on them that work in full simulation. However, place these templates on blank screens with no other vessels or reactors. Functionally everything will work, but because these graphics do not resemble anything in their process, the operators will not associate them with the final product.

Insight: Many a young person has gone to an interview wearing ragged clothes, uncut hair, and an attitude of "If they judge me by my outside appearance, then I don't want to work for them." As an interesting aside, very few (if any) get hired.

Rule of Thumb: There is a tremendous amount of wisdom in the expression, "You never get a second chance to make a first impression." Make the effort to make sure that the first impression at an interview, or of a system review, or of a control panel is the absolute best it can be.

> **"... the first impression you make will likely stick for a long time."**

Tip #31: Project Management Is Really Risk Management

Hunter Vegas (Project Management/Execution)

As a project manager I am responsible for a myriad of activities, but if I consider them in aggregate I realize that they are just different manifestations of the same thing: my job as a project manager is to recognize risk and manage it.

Concept: Successful project managers are masters at evaluating, mitigating, and avoiding risk. Their job is to bring a project in on time and within budget. Anything that can jeopardize that is a threat and must be avoided.

Details: The best project managers recognize problems BEFORE they are problems and either find ways to avoid them or have contingency plans in place to effortlessly address issues as they occur. They must maintain a vigilant guard against problems (or potential problems) and mentally have a Plan B (or Plan C or even Plan D) available to immediately bring to bear if necessary. Jiminy Cricket makes very poor project plans – wishing upon a star and hoping that things work out is NOT the way to run a project. During the project, continuously monitor how things are going against the plan and adjust quickly as necessary. If a manager acts when the project is a third of the way through, then he or she has time to catch up and still make the budget and/or schedule. If the manager delays taking mitigating action until the project is 80% or 90% complete, then failure is virtually assured.

Here are some potential sources of risk and ways to eliminate them:

- Choose the project team wisely. If at all possible, try to keep the same core team and only add a new member on occasion. A core team knows each other's strengths, weaknesses, and quirks and can communicate at a much higher level with very little effort. They also can bring to bear a strong history of techniques from past projects and lessons learned.
- Know the technology. If the team is new to a DCS or network technology, try to work some smaller projects and gain some experience before jumping into a larger project.
- Know the process and/or the client. Understanding the process and the client's needs is critical for project success. If the process is untested or the client is new to the team, budget more time to get to know the client and understand his needs.
- Send graphics or partially completed software to the client for early review. If possible, develop a graphic and all of the underlying modules (with imbedded simulation) as a unit and forward it to the client for review upon completion. If there is a disconnect between the client's expectations and the team's work product, the team needs to know that early!
- Build and test the control panels in advance. (This was discussed previously in Tip #29.)
- Avoid the latest version of software. If the project does not require the latest functionality, do not rush to utilize the latest software release. With increasing frequency, ven-

dors are releasing poorly tested software and letting their customers debug it for them. Let someone else fight that battle.

Watch-Outs: Order any project material early and constantly check on its progress once ordered. Many projects have been delayed because the equipment failed to arrive on time. Sometimes the purchase order never got placed by the Purchasing Department, sometimes the vendor lost the order, and sometimes the factory got behind and failed to deliver or delivered the wrong thing. Assign a person to check up on each purchase order and when items arrive, open the boxes and make certain the material is correct.

Exceptions: There really are not any exceptions for risk management. EVERYTHING can be a source of risk and should be monitored. A subcontractor can have a long track record of success but fail to deliver on time due to internal organizational changes or workload. Valued vendors may have manufacturing problems and the usual two week delivery may extend to 12 weeks. Take nothing for granted.

Insight: If the project involves programming a safety interlock PLC consider staying at LEAST one major revision behind and if that revision was short lived or had problems, it may even be worth staying TWO revisions behind. When dealing with safety interlocks, the programming is rarely complex and system stability is far more important than getting the latest bells and whistles (and the bugs often associated with them).

Rule of Thumb: As a project manager, continuously monitor everything for signs of problems and immediately inquire if something seems amiss. No project manager wants to be surprised.

Tip #32: General Project Management Tips

Hunter Vegas (Project Management/Execution)

Over the years I have learned a number of valuable tips, either the hard way or from my talented colleagues. Here are some points that will be extremely useful to any project manager.

Concept: Successful project management requires a diverse skill set and covers a broad range of activities. Most project managers acquire that skill set the hard way, but hopefully this list of tips will help you acquire those abilities in a much quicker and less painful fashion.

Details: As mentioned previously, a good project manager is really a risk manager and a successful project manager is constantly working to identify and eliminate risk to the project. Beyond that, here are several other key concepts that a good project manager should understand.

- Good, fast, cheap – pick any two. This is an age old project management maxim, and the simple fact is that you cannot have it all. If the project must be completed quickly, it will either suffer in quality or cost more. If project expense is paramount, prepare to sacrifice quality or speed of execution. And finally, if the project must be of the highest quality, then it will either cost much more or take longer to install. Manage the client's expectations. If the client is making unreasonable requests and demanding a fast track, high quality, low cost project then consider either declining the project entirely or informing them at the onset of the impossibility of satisfying their requests.
- Time nearly always trumps cost. Most projects have a one or two year payback. Therefore, a one week delay is usually worth 1 – 2% of the project cost. Spending a bit more to save time is usually justified by the economics.
- Never commit to something that you cannot deliver, but ALWAYS deliver on those items to which you commit. Follow this advice and your team will quickly gain the respect of your co-workers and clients. Instill this concept in your team and success is virtually assured.
- Never low-ball or undercut an estimate. Returning excess funds is infinitely easier (and much better career-wise) than over-running a job budget and begging for more funding.
- If the client wants to cut project cost, cut the scope rather than cutting out the contingency funds. NEVER assume everything will go perfectly – because it will not.
- Plan the work and work the plan. Take the time to plan the project at the start and continually compare the actual project status to the plan. Find simple metrics that will provide a quick and accurate picture of the activities and highlight possible problems early. Fixing a problem is difficult if you don't know it exists!
- Know the team. Pick team members whose talents and quality are known and whose trustworthiness is a given. If their "first time right" percentage is low or they are new to the team, cross check their work early in the process.

- Realize that there is rarely one path to accomplish a goal. As a project leader it is easy to become so focused on proving your own approach that other viable options may be dismissed. Budget and timing limitations may dictate that things be done in a certain way, but if possible let the people under you run with their ideas. This makes them stronger team members and they will often become more engaged and work harder to prove that their concept will work.
- Communicate, communicate, communicate – be sure the team understands their role and understands the project goals. Formal meetings are not required and can often be replaced by constant informal communications that keep all of the team members on the same page.
- Identify scope changes early and document them well. If the contract is a fixed price contract, get formal approval BEFORE doing the work. When the client is desperate to get something in, he will say anything to get it done. However, people tend to forget those promises when the money gets tight at the end of project and they are presented with a verbally approved scope change for final sign off.

Watch-Outs: While communication is extremely important, realize that the productivity of meetings varies with the inverse of the square of the number of attendees. A two or three person meeting is highly efficient. Productivity begins falling precipitously after five attendees and approaches zero as the number grows above ten. Short, informal meetings of a couple of people will usually provide the best results.

Exceptions: Working the plan is important, but even more important is the ability to realize that the plan needs to be able to change! Be flexible. If things are not going as they should, do not be afraid to change direction early and quickly to rectify problems.

Insight: Young engineers often see themselves as managers before they are ready for the role. To be successful a manager has to have the skill and desire to manage clients, motivate and develop others, provide technical direction, ensure quality, keep the schedule, watch the budget, mitigate risk, do the billing, etc. If an engineer is missing some of those skills he or she should consciously pick the weakest one and work to improve it.

Rule of Thumb: The best project managers probably did not set out to BE project managers. In most cases they started out as engineers who over time proved their ability to successfully run increasingly large portions of projects and manage others. Most make the job look easy, but the role is a lot more difficult than it might appear.

> **"Good, fast, or cheap ... pick any two."**

Tip #33: High Pay Is Great but Life Is Too Short to Stay in a Job You Dislike

Hunter Vegas (Career)

Around the middle of my career, I was becoming disenchanted with the reorganization of my plant and decided to pursue another job. I found what appeared to be an excellent opportunity – big raise, generous relocation package, awesome benefits, and a lucrative bonus. It sounded like a "can't lose" opportunity and I jumped at the chance. Unfortunately, it was NOT at all what I expected.

Concept: Getting a high salary and/or bonus is a great thing, but if an engineer hates his or her job, the money is not worth the pain.

Details: Everyone likes to be paid well. In this materialistic world, a large paycheck is a badge of success and having a high standard of living is certainly a nice perk. However, a large salary is not worth the price if it demands misery at work. If you find yourself waking up each morning and dreading the thought of returning to your job, the time for change is at hand.

When you are looking for a new job, you should "interview" the company as hard as the company is interviewing you. Obviously, company personnel need to ask enough questions to make sure your skill set meets the needs of the position, but it is just as critical that you make certain the company is a place where you can contribute and enjoy working. Domineering bosses, militant unions, and bad managerial/worker relations make for a terrible work environment. You should also keep an eye out for "dead end" positions that offer no growth or learning opportunities. If there are any options at all, strongly pursue alternate job possibilities.

Consider taking future co-workers out to lunch to get them to talk and/or see what they might be like. If the company is local, find an ex-employee and see why the employee left. If you know some of the company's clients, see how they feel about the company. All of these can be great sources of information.

Watch-Outs: When you are interviewing, be sure to talk to a wide enough range of people. If you only interview within a small group, the politics and/or issues of the rest of the organization may not be apparent. During your tour, talk to operators, technicians and supervisors. Strike up casual conversations with administrative and office personnel and get business cards from any individuals who seem most open. An evening call to those individuals may be illuminating.

Exceptions: If you are out of a job and have bills to pay, ANY employer looks like a good option. However, even in those circumstances you should try to be as selective as possible.

Insight: As unorthodox as it might seem, it may be possible to "test drive" a prospective employer if they are local. If a company seems to be a good prospect, offer to take a week or two of vacation from your current job and work for the new company with the understanding that if either is unhappy with the arrangement, both can go their separate ways no questions asked. However, if both are satisfied after two weeks, you can resign from your old job and accept a position with the new company. If there are no non-compete/legal concerns, it may be possible to arrange to get paid for those two weeks after signing on with the new company.

Rule of Thumb: Life is too short to hate your job. A good paycheck is nice, but if you find yourself dreading the return to work each morning or if your job requires you to be on medication (including alcohol), it is time to seriously consider a change.

"A large salary is not worth the price if it demands misery at work."

Tip #34: Take Time to Learn Exactly How Instruments Work

Hunter Vegas (Career)

This tip is actually a compilation of Tip #1 "Never Stop Learning" and Greg's tip on "Seek Principles," Tip #66. As a young engineer I spent a great deal of time and effort learning how instrumentation worked. Once I had a thorough understanding of that information, I could evaluate competing instrumentation designs and better select the best choice for a given application.

Concept: Making the effort to study and understand exactly HOW instruments work will generate a lifetime of benefits. This should be a high priority for any young automation engineer.

Details: Young engineers are often amazed by the ability of a seasoned engineer to ask a handful of pointed questions and immediately select the best instrument type and manufacturer for a given application. This skill does not require a photographic memory nor 54 years of instrumentation specification experience. It simply requires that the engineer understand exactly HOW the equipment functions. Once you know that, it is a straightforward effort to systematically eliminate instrument types that cannot work and gradually pare the number of options to a select few. A few more questions will often reveal the superior option. If you know the different designs offered by the various vendors and know the strengths and weaknesses of those designs, you can quickly name the particular make and model best suited for the application being considered.

Learning all of that information is not a minor undertaking, but it is not monumental either. When an opportunity arises, take the time to learn about unfamiliar instrumentation and seek to understand the details behind its function.

Watch-Outs: The pursuit of instrumentation knowledge should begin with unbiased sources (such as ISA or other technical organizations). These organizations tend to provide a more complete picture of instrument design and list the pros and cons of each technology. Vendors often give a more limited view of the technology, highlighting their own product line but downplaying (or even ignoring) competing designs. Vendor information can provide useful information about the particular design details of different instruments, but it is a poor place to start if you are just entering the field.

Exceptions: None.

Insight: Spending some time in the field with the maintenance technicians can be invaluable to a new engineer. If you develop a good relationship and rapport with the techs, you will learn a great deal about the installation, repair, and troubleshooting of various instruments. Knowing how instruments can fail is as important as knowing how they work.

Rule of Thumb: Take a few basic ISA instrument courses, read some books, and ask co-workers about instrumentation and how it works. A detailed understanding of these devices is critical for any automation professional.

EXCEPTIONS: None.

Tip #35: Operators Can Make or Break an Automation Project

Hunter Vegas (Career)

Senior management holds the purse strings, engineers do the design, and Production/Operations supervisors dictate what they want, but the operator (who will use the system 99.9% of the time) rarely gets much involvement in the design of a control system. However, that group, above all others, will ultimately determine whether the project is a success or a failure.

Concept: The operators are critical to an automation project's success. They must be engaged early and often or a negative outcome is highly probable.

Details: Over the course of a typical automation project, engineering and production management are in constant contact, refining the technical details and sorting out operability requirements. The system is designed, fabricated, installed, and tested and then promptly tossed to the operators to use from that point forward. This operational transition period tends to be overlooked or minimized, yet it is absolutely critical to the ultimate success of the project. If the operators hate the new system or are poorly trained on it, failure is highly likely.

Operators use the system every hour of every day, yet they are often minimally involved in the design of the system. Sometimes the Management or Production staff intentionally decides to exclude them, sometimes the operators are offered an opportunity to participate but are too busy or uninterested to be bothered. Regardless of the reason, the fact remains that if the operators are NOT involved, project risk escalates dramatically.

It is the automation engineer's job to GET the operators involved by whatever means it takes. As a minimum get one or two of the more experienced operators to review graphics as they are produced so they are continuously involved in the process. As new graphics and modules are completed (with imbedded simulation), release them to the operators for review and/or training. Invite at least two of the most respected operators to the Factory Acceptance Test (FAT). If they ask for changes that are easily made, give them what they ask for. If they ask for a difficult change, explain why that particular change is difficult and offer alternatives that may accomplish the same goal. Strive to make the operators feel a part of the process and that their requests are being heard. Buy-in is crucial.

Watch-Outs: In these days of lean staffing, plant operators simply may not have the time to review a new system even if it is available. This can be a difficult situation, but it can be rectified if the plant is willing to pay the operators overtime to spend an hour or two on the system before/after their shift. Probably the best way to encourage this behavior is to have a "test" associated with the new system that must be passed prior to allowing the operators onto the new system.

Exceptions: If the plant and/or process are new, the operators may not have been hired or may not yet be officially assigned to the plant. In this case it may be possible to arrange operator reviews of the graphics as a training exercise (assuming they are available).

Insight: Many control systems make it easy to create low level simulations of the process. It does not require much programming effort to allow valves to open/close, pumps to start/stop, PID control loops to hold a setpoint, and levels and flows to realistically change. Even the simplest simulation allows a thorough testing of a complex control system and is often a great tool for operator training (now and in the future).

Rule of Thumb: Get the operators involved in the design process early and listen to what they say. They will use the system 99.9% of the time and their buy-in and ultimate approval are critical to an automation project's success.

> **"It is the automation engineer's job to GET the operators involved by whatever means it takes."**

Tip #36: A College Degree Does NOT Buy You Respect, Nor Is It a Good Indication of What You Know

Hunter Vegas (Career)

The corollary is also true – the *lack* of a college degree is not a reliable indicator of a person's level of knowledge. I have seen young engineers who expected (and even demanded) respect and deference even though they had done nothing to earn it other than graduating. I can also say that some of the brightest and most knowledgeable mentors I have known never had a degree of any kind. Respect is never awarded – it is earned. (See also Tip #7.)

Concept: A college degree is proof that a person attended a bunch of classes and passed some exams. While that person may know a great deal of theory, he or she may know little that is useful to an operating plant. Do not automatically assume that a college degree means a person is exceptionally knowledgeable, and never undervalue a person who lacks a degree.

Details: After enduring years of labs, dull professors, and incomprehensible calculus and thermodynamic classes, a young engineer is rightfully proud of his or her degree. Unfortunately, some engineers feel that the new diploma automatically means they are worthy of a higher level of respect and worse, they begin to look down on their non-degreed co-workers. This is NOT a promising start to a successful automation career!

People may be bestowed with titles but respect is always earned. Even if an engineer is young and inexperienced, he or she can quickly gain the respect of co-workers by respecting others, being eager to learn and help, and by listening. Alternately, an engineer can be experienced and knowledgeable, yet earn no respect due to a condescending attitude and a tendency to belittle others. Treat everyone with respect, regardless of their station in life, and seek to earn their respect through your actions. This simple concept will serve you well throughout your career and life.

Watch-Outs: As stated in Tip #7, it is probably best to downplay your advanced degrees and titles unless specifically asked about them.

Exceptions: None.

Insight: Some of the most innovative and interesting information and ideas you will encounter in life will be from people outside of the automation engineering profession. Operators, technicians, supervisors, etc. all know the plant far better than you do. Talk to them, ask their advice, share your ideas with them, and get to know them. You will find the effort well worth your time. When a great idea does come from the ranks be sure that they get credit; otherwise that may be the last idea they ever feed you.

Rule of Thumb: You must earn respect, and show respect to others. An engineering degree (or lack thereof) is a poor indicator of an individual's knowledge level.

> **"Operators, technicians, supervisors all know the plant far better than you do."**

Tip #37: Get Past the References You Are Offered

Hunter Vegas (Career)

Over the course of my career I have never been given a reference that provided anything but glowing reports and testimonials. Perhaps your experience will be different than mine, but the odds are slim. In every case each reference I contacted gushed ad infinitum about how wonderful the candidate was or how positively flawless the new product was, etc. From this you might conclude that it is a waste of time to even bother to contact the references, but this would be a bad assumption. The real reason to contact the references is not to hear the glowing reports that you are bound to get – it is to get the names of the second tier of people who CAN give you an unbiased view of the person/product you are investigating.

Concept: Nobody (not interview candidates nor control system or equipment vendors) will offer a bad reference. (Of course if they DID offer a bad reference, that would make the selection process much easier!) However, you can use that contact to get names and leads on other people who might provide a more accurate assessment.

Details: References that are provided as part of any evaluation process are practically worthless. No interview candidate and/or company would be so foolish as to provide references who will mention anything but complimentary details. However, that list can provide a path to a more complete evaluation with a bit of investigative work. If you are checking an interview candidate's references, ask for the names of co-workers, bosses, or other individuals who might have directed the employee or worked with them. If possible, contact these individuals and see what information they have to offer (or what other people they know). Keep in mind that most people will speak more openly in a face to face conversation than over the phone. They will also tell a lot more to an acquaintance than to someone they do not know. Take advantage of this and try to meet people face to face or ask around and find if "someone knows someone" who can access these people.

Evaluating a company or a product is easier. If you are evaluating a product (such as a control system), ask for a list of recently installed systems. If you are evaluating a company (such as an engineering firm or a system integrator) ask for a list of recent clients. In either case, use the initial list of contacts to work deeper into the organization and access a larger group who might not have been primed by the sales person. The initial contact will often be a manager or other high level administrator who is far removed from the day to day issues and is not aware of any problems. Even worse, the initial contact may have been responsible for selecting the company and/or product and is loath to admit any problems that might reflect poorly on their career. However, if you can get past that initial contact and get the names and contact information of other engineers, supervisors, and operators, you may find a host of individuals who know a great deal and are more likely to share the pros and cons of the company or product being considered.

Watch-Outs: The Human Resources department is probably the LEAST suitable source for information about an interview candidate. When a Human Resources person identifies themselves on a phone call, the person on the other end usually clams up and will volunteer very little. Similarly, if a Human Resources person is asked about a particular candidate, they can offer very limited data or they risk making themselves the target of a lawsuit. Networking contacts (outside of Human Resources) will yield much more information.

Exceptions: An interview candidate is often currently working for a company and does not want that company to know they are job hunting. In this case it may not be possible to contact references at that company, much less get past them to dig deeper. One option is to ask for references and vet the employee's work at a previous company (assuming they worked at another company before the one where they are employed today). Another option is to ask the candidate for the name of a trusted co-worker who might shed some light on the candidate's qualifications without compromising the secret. Either way, the available information will be somewhat limited.

Insight: When you are evaluating a control system it is always best to go on site if at all possible. If the tour has been arranged by the vendor, they will likely try to limit contact to a few carefully chosen people. Ask for a control room tour and during this time have the evaluating team "fan out" to talk to technicians, operators, supervisors, etc. Ask them about the system and about any individuals who might know more about the system "behind the scenes." These casual conversations and contacts with other people will provide the evaluation team a better picture of the strengths and weaknesses of the system in question.

Rule of Thumb: Never be satisfied with the reference provided. Treat it as a first step toward engaging a wider group of people who can actually provide a more realistic picture of the person/product/company being evaluated.

> **"Nobody ... will offer a bad reference."**

Tip #38: Derail Scripted Demonstrations

Hunter Vegas (Career)

We have all been there. The sales person stands before the crowd demonstrating a control system. He clicks here and lays down a pump, he clicks there and adds a control valve, he opens this window and everything magically links and is fully functional. The system is a flawless masterpiece. Nothing can go wrong … go wrong … go wrong … Anyone who knows me knows that I cannot stand scripted demonstrations. They are like the references discussed in the previous tip – the demonstration will always be flawless. I want to see what happens when I throw the system a curve!

My favorite war story about this particular subject happened in the late 1990s. I was on a team evaluating two brands of expensive simulation software that was to be used for modeling and optimizing our large continuous plants. The plant manager was friendly with one of the salesmen and let it be known which product we were to choose. However, the team quickly realized that the product he had selected was far inferior to the other option. Given the manager's personality, there was little chance that the team's opinion would be allowed to override his own.

The team arranged for each company to give a presentation of their product. The company favored by the team presented first. The product worked well and was quite adaptable. The sales person did not follow a script, but asked for ideas from the audience and built a process model before the whole group. The plant manager's favorite company gave its presentation a few days later. The salesman knew that the previous presentation had gone well so he chose to use a recently released beta version of his product, which had a slick graphics package and lots of bells and whistles. During his demonstration he clearly followed a script … adding a specific column type here, a certain pump there, etc. I let him continue for about a half hour – leading us down a path that had been carefully choreographed to avoid any problems.

Then I raised my hand and asked an innocent question – something about the system's ability to handle a different configuration. Perhaps the column had a recycle stream coming in 1/3 of the way down the column or used a different control philosophy than the one he had chosen. "Of course," he answered, "you would just add those elements to the model."

"Great," I responded. "Do you mind showing us?"

He blanched but had no choice. Three mouse clicks later the system sputtered and blue screened. At that point the salesman had to admit that everything he had shown us was a beta version of software that was not fully tested. Needless to say the team selected the other product and the plant manager had no option but to back the team's decision.

Concept: Like the references in the previous tip, scripted demonstrations are practically worthless. Let them get started and then start asking questions that force the sales person to depart from the script. If the system is good, it will easily handle the diversion and perform well. If the system is designed poorly, it will quickly become clear.

Details: Regardless of the product being demonstrated, most sales people like to follow a well-choreographed, thoroughly tested script. The risk is low and the script has been specifically designed to highlight the strengths and minimize the weaknesses of the particular product. The sales person's job is to make sure everything goes to plan. However, the audience members WANT to know the product's weaknesses and flaws. They have a vested interest in forcing that demonstration off its script so that a more complete evaluation is possible. When attending a demonstration, the audience must recognize that the demo is scripted and take pains to force the sales person in a new direction.

Recognizing a scripted pattern is usually easy. The sales person appears to be following a predefined sequence and is not eager to deviate from the path. He or she might even be referring to notes as they proceed. In such cases, ask about a different feature and then ask how easy that feature is to implement. Once the sales person answers, it should not be much of a problem for him or her to DEMONSTRATE the feature to the crowd. At this point the sales person is trapped, and must acquiesce or start providing reasons why they cannot.

If possible, ask for an opportunity to "drive" the system and take part in the demonstration rather than simply watching it. Now you are in control and can easily go down different paths and try different features to better evaluate the system's strengths and weaknesses. If the system is good then it should have no problems handling whatever you ask it to do.

Watch-Outs: Sometimes a presentation is being made in front of a large group of diverse customers and your question is not of interest to most of the people in the room. Rather than taking up everyone's time chasing a side issue, take it up with the sales person on an individual basis after the demonstration is finished.

Exceptions: Occasionally a product is so bad or the sales person is so nervous that the scripted presentation is all they can do. In such a case, there is nothing gained by embarrassing the person any further by forcing them off the script. Just recognize the situation for what it is and move on.

Insight: This same idea applies to factory acceptance tests. Do not let the integrator take the system through a scripted test sequence to prove its functionality. Instead, intentionally close the wrong valve or press the wrong button. Force the system off the beaten path. It is certain that the operators will do that eventually and it is far better to discover problems before the system is on line! If possible, require that the integrator be completely hands off during testing to prevent them from nudging the system along during testing.

Rule of Thumb: A well designed system should be able to handle whatever is thrown at it, especially the "non-standard" situations. Every now and again step off the beaten path, throw the system a curve, and see what transpires.

Tip #39: Know the Codes

Hunter Vegas (Career)

Needless to say, studying the National Fire Protection Association (NFPA) Codes and National Electric Code (NEC) is a wonderful cure for insomnia but they (and others like them) are important! Possibly the most useful course I have taken was a preparation course for an electrical contractor's license. I actually took the course because I needed some PE credit hours, but after hearing how difficult the North Carolina test was and given my newfound understanding of HOW the code was put together, I decided to purchase the supplementary training materials and pursue the license. The test was distressingly difficult – essentially memorizing the entire National Electric Code – but the information has been invaluable on countless occasions. The codes are long and arduous to read, but correctly applied they can save people's lives and prevent millions of dollars of equipment losses.

Concept: Reading and knowing national codes and technical standards is a challenging, but necessary task for a higher level automation professional. (For simplicity, "codes" means both codes and standards in this Tip.)

Details: Engineers graduate from school and are thrilled with the prospect of having their studying behind them – but their effort in school is only the beginning. Most engineers will be studying codes and/or code updates for their entire career. The ones who do NOT know these codes will find their career stagnating quickly.

Do NOT start this process by grabbing the nearest code book and trying to read it. Under such conditions the average engineer will either quit after a few days or soldier through it, never understanding the information or recalling any of what they read. Instead, try this process:

- Select a code that is the most useful to your current job description. It might be the S88 Batch standard, the NEC, or one of the NFPA, UL, or ASME codes or ISA Standards.
- Once the code is selected, try to find a short class and/or book that explains it and its application. This information will be important as you begin reading.
- NOW read the code (or at least sections of it). Start with the parts that are most applicable to your present job description.

For example, in the specific case of S88, consider reading an introductory book such as *Applying S88: Batch Control from a User's Perspective*[1] about two guys creating a batch control system for an ice cream plant. The information in that book is easy to read and explains the application of the standard in an actual plant. (The book makes great sense; the S88 standard itself is difficult to follow.)

1. Applying S88: Batch Control from a User's Perspective, Jim Parshall and L.B. Lamb, ISA 2000.

If you are learning the National Electric Code, take a preparatory course for a state electrical contractor's license. Even if you have no plans to actually get a license, the class will explain the often mystifying arrangement of the NEC and will help you quickly locate information. Do NOT try to read and/or memorize the code cover to cover. It is huge and covers a wide variety of specific situations that have little bearing on the average automation engineer's job. DO read the early generic chapters on branch circuits, grounding, and conductors as well as the chapters on motors, and read the sections on hazardous area classifications if applicable to your situation.

Make an effort to pick a code and/or code section and learn a new one every year. Most engineers will find the following codes particularly useful:

- ISA-88 Batch
- ISA84 Safety Interlock Systems
- National Electric Code (certain sections)
- NFPA 85/86 (Boiler/Furnace controls)

There are countless others, each with its own purpose and application.

Watch-Outs: Beware of code "evangelists." Occasionally you will encounter an individual who quotes codes by chapter and verse and uses this tactic to bend other, less knowledgeable people to their will. Some of these people ARE that knowledgeable, but others really do not know as much as they pretend. (Electrical contractors will occasionally use this technique to get lucrative change orders.) If a person demands that a major work scope change must be implemented to "satisfy the code," it is worth getting a second opinion from a knowledgeable, but independent third party.

Exceptions: There is really no excuse to avoid learning codes. Do not rely on the knowledge of others to know and/or interpret a code. Most of the codes have gray areas instead of being strictly "black and white" and have "outs" that may provide alternate options for certain situations.

Insight: There are not many engineers with commanding knowledge of the codes, so they are often well paid and valued by a company.

Rule of Thumb: Try to learn at least one relevant code or one code section each year and seek to apply that knowledge often so that it is not lost.

Tip #40: Engineers Are Generally AWFUL Communicators. Do Not Fall into That Trap.

Hunter Vegas (Career)

Consider the stereotype of the engineer. He (or she) is shy, socially inept, and rarely the life of the party. Understanding Fourier transforms and partial differential equations is no problem but standing in front of an audience and giving a presentation is a nearly insurmountable challenge. If he does give a talk he tends to mumble and look away from the audience when he speaks, puts WAY too much information in small type on his PowerPoint slides, and tends to read the slides to the audience as he drones on.

...sound a bit too familiar?

Unfortunately, that description is close to the truth for many engineers, and often a more technically qualified engineer has been passed over for a promotion by a candidate with less experience and knowledge but with better communication and presentation skills.

As a high school student I realized I was headed down this road. My math scores were excellent but my English/verbal scores were at best mediocre. I was struggling in a writing class where the instructor held me to a much higher standard because he KNEW I was capable of it – but I was not delivering. I finally decided it was time to address the problem, knowing it would only worsen with time. I sought help with writing and joined the speech and debate team, where I signed up for extemporaneous speaking. (You pick a topic out a jar then have 15 or 20 minutes to prepare and deliver a five minute speech.) It forced me to organize my ideas and think quickly on my feet – useful skills, to be sure.

Later in life I took courses on media presentation as part of my Hazmat team training as well as a few sales presentation classes. Such training has been invaluable to me over the years.

Concept: A knowledgeable engineer is worth a great deal to a company. However, a knowledgeable engineer who can write and present is worth much, much more!

Details: The average engineering-bound high school student has a profoundly skewed math/verbal aptitude ratio. Math and science classes are no problem but most engineering-bound students struggle in English and communication. Unfortunately, this inability to communicate will hamstring them throughout their career. People will steal their ideas because they cannot sell them to upper management. Many will lose promotions to less technically qualified applicants because the other candidates speak and "show" well. Opportunities for exposure to upper management will be lost because they look so uncomfortable in presentations. However, a highly qualified engineer who can also write and present is a rare thing indeed. The combination of strong technical talent and excellent communication skills is extremely valuable in most organizations.

If you are still in school, do not shy away from writing and speaking courses – TAKE THEM. If you are already in the work force, take a few writing and presentation courses and sign up to give a technical presentation at your local technical society meeting or national conference. Seek opportunities to write or present on the job.

If you are scheduled to present, take the time to practice and more important, record your practice talk on video. It may be painful at first to watch yourself speaking but you cannot fix a problem until you experience it first-hand.

Watch-Outs: Engineers often assume that their audience has the same level of technical knowledge as they do and is as excited about the subject as they are. However, both assumptions are usually false. Some, or perhaps many, in the audience do not know all of the technical background and will get lost quickly. Combine that with a poorly presented or dull topic and the snoring can be intense. Take the time to quickly explain the technical background concepts well enough to ensure that everyone is on the same level, then proceed with the new information. Most importantly, make the slides INTERESTING! Do not use dull PowerPoint slides with tiny fonts, no graphics, and line after line of bulleted items. Instead, use interesting pictures, simple phrases, and large, easy to read fonts to excite the audience and entice them to listen to you.

Exceptions: If you already have your own TV or radio talk show, then you are probably exempt from having to pursue this tip further.

Insight: When you are giving a technical presentation, it may be worth taking the time to create two versions of the slides. Make up a set showing all of the technical points and details and provide that version to the audience as a handout, then create a second set that has the major talking points but is much more visually appealing, with fewer words and lots of pictures and/or graphics. The audience will appreciate and enjoy the presentation (which will be refreshingly different than MOST of the presentations they have seen) and they will have the technical handout for reference and notes.

Rule of Thumb: Many engineers are not naturally adept at writing and/or presentations, and they pay a price for that failing throughout their career. Like anything else worth doing, good writing and presentation skills take training and practice but the effort WILL be rewarded.

> **“... a highly qualified engineer who can also write and present is a rare thing indeed.”**

Tip #41: If Possible, Get Plant Experience Early

Hunter Vegas (Career)

You're studying the elephant in school. You look at the pictures, learn about the anatomy and physiology of the animal, research its habits, etc. With enough study you might write a paper or two, and in time you might even be considered an expert on the beast.

Studying an elephant is one thing, but it is quite another to see one up close. The toughness of its skin, the enormous size of the creature, and the deafening blast of its call are impossible to adequately describe in a book. Theory is nice (and necessary), but reality is infinitely more educational.

In a similar vein, control theory is often taught in school. Processes are mathematically simulated and control algorithms are developed. Many of these algorithms provide significantly "improved" control response in their simulated world and are a frequent subject of research papers. Some academic professors consider themselves an expert in the field but have never actually CONTROLLED a process in real life.

Concept: The most versatile and sought-after engineers have worked in a plant and have been involved in day-to-day plant operations.

Details: If at all possible, seek out a position in an operating plant upon graduation from college. What is more, seek opportunities to work side by side with the operators and technicians as soon as possible. Exposure to the real world of controls instead of the theoretical concepts taught in school will be illuminating. It is one thing to talk about installing instrumentation in accessible locations and planning for maintainability, but it is quite another to be chewed out by a technician because you had a transmitter mounted six feet outside the hand rail! Simple things like installing drain and isolation valves or positioning frequently operated valves low for easy operator access are almost never taught in class, yet these modifications mean everything to the operators and technicians who must run and maintain the plant. A couple of years of plant experience will provide new insight into the minor design changes that can make the difference between an easy plant to run and one that is a constant battle.

Some new engineers immediately pursue jobs at engineering firms, but this experience is not nearly as useful as operating plant experience. Most contract engineers design a plant, start it up, and leave. They never get the feedback from operations personnel on what worked or failed to work, or worked well initially then failed over time. It is much more difficult to learn about the realities of the manufacturing plant in such a firm.

Watch-Outs: This same concept can be invaluable to seasoned engineers as well. If you are evaluating outside engineering firms, beware of firms that are staffed with principal engineers with no plant experience. Such engineers may be technically adept, but they may have no clue about how to design a plant for efficient operation. In addition, ask for and study the resumes of the engineers who will be assigned to YOUR project and do not let the firm "bait and switch." Some firms will use the top engineers to sell a job and then assign weaker engineers to do the work.

Exceptions: Sometimes the recently graduated engineer cannot find a job in a plant and the only option may be a consulting engineering house. If that describes you, volunteer for start-up opportunities and ask to be brought along with the experienced principal engineers during sales calls or HAZOPs. Seek to gain as much in-plant experience as possible.

Insight: When searching for a job, seek out plants that have at least one seasoned engineer under whom you can work. Such a situation will allow you to learn much quicker, especially if the engineer is much older and does not consider you a "threat."

Rule of Thumb: Upon graduating, seriously consider a position in a plant environment. The hours can be longer and the work more difficult, but the learning opportunities are worth the effort.

"Theory is nice, but reality is infinitely more educational."

Tip #42: Money Drives Corporate Behavior

Hunter Vegas (Career)

If you are the optimistic sort who always sees the good in your fellow man and always takes people and companies at their word – skip to Tip #43 and save yourself the angst. However, if you are like most engineers and tend to be a bit more skeptical and jaded, then this tip is for you.

Political correctness has never been my strong suit and for better or worse I call things as I see them. If I spot a gorilla in the room, I am usually the first to point it out so that everyone can quickly get past that distraction and work to address the issue at hand. I firmly believe that you cannot begin to resolve a problem until everyone agrees that a problem exists.

In that spirit let me introduce this tip...

Concept: While nearly every company trumpets People, Safety, and Environment as a higher priority than profitability, the fact is that MOST corporate decisions are driven by the bottom line.

Details: This concept flies in the face of practically every corporate sales and marketing campaign in existence. Their advertisements would have us believe that chemical companies champion the environment above all else, that the core value of pharmaceutical firms is the elimination of human suffering, and that all companies consider their employees to be their number one resource. This is all wonderful and fine, except that it is rarely true.

All companies champion money. They have to. No company can continue to operate while constantly losing money. Short term revenue-negative decisions may be made, but ultimately the company must make a profit if it is to survive. All things being equal, a company will act to improve its profit.

So are safety and the environment REALLY a priority? Yes, because injuring people and getting caught dumping hazardous chemicals can be expensive. (BP certainly discovered this in the Gulf of Mexico in 2010!) However, if a company can cut corners, bypass safety interlocks, and modify procedures to get a unit in production quicker and the risk is deemed acceptable, then there is every likelihood that that is exactly what will happen. (BP and their affiliated companies are being accused of those very activities in the courts today.) So what is the point of this tip (other than to depress people)?

If you as an engineer understand this dynamic, you can use it to encourage the company to do the right thing if only because it makes financial sense. If the company is flouting (that is, willfully disregarding) a safety procedure or bending an environmental regulation, calling attention to it may fall on deaf ears. However, if you can quantify the financial loss of spilled product or lost production due to an EPA restraining order, suddenly the issue has significant financial ramifications and may be treated differently. Similarly, it may benefit the production numbers to allow a waste stream to be generated even if the environment is impacted. However, if the actual cost of the waste stream can be quantified and highlighted, it may behoove the company to reconsider their position. (See next tip for details.) In short, when you are faced with a situation where your company is acting in ways it should not, pull together a complete list of reasons to stop that activity before you approach management. Your list should include safety reasons, environmental reasons, and most importantly, financial reasons. With such a list, your odds of redirecting the company's actions will be greatly enhanced.

Watch-Outs: If your company seriously flouts a safety or environmental regulation and people's lives are at stake, call immediate attention to it regardless of the financial impact. Start with your direct supervisor and if you are ignored, inform him or her that you are going to the next level and immediately do it. If you get no satisfaction at that level, continue until the issue has been resolved. Exhaust all levels within the company before turning to outside entities (such as the police, OSHA, EPA, etc.) Doing the right thing is not easy but living with the deaths of a couple of co-workers because you failed to act is not a walk in the park either.

Exceptions: Occasionally a company REALLY WILL do the right thing regardless of the cost. This is great to see and it is certainly a company worth working for. Unfortunately, this seems to happen much too infrequently, especially during hard economic times. People usually start off meaning well but the pressure to improve the bottom line tends to eliminate this type of behavior over time.

Insight: When calculating the cost of a project, a decision, or some other activity, do not fail to consider the intangible costs. Spilled chemicals may not cost the company much from a raw material perspective, but having a TV reporter broadcast the fact on national news certainly IS expensive. Similarly, eliminating a waste stream may not save much money in material costs, but if the waste reduction drops the company significantly lower on the EPA's list of top polluters, the idea may seem much more attractive. Take the time to look at the bigger picture and determine other benefits that the company may not have considered.

Rule of Thumb: Follow the money. Historically, it has been an excellent predictor of corporate behavior.

> **"No company can continue to operate while constantly losing money."**

Tip #43: Small, Continuous Waste Streams = BIG Money

Hunter Vegas (Career)

When an engineer starts his or her first job, he is often looking to make a name for himself in the plant. Saving the company big money and getting an award and/or recognition is a pretty good place to start. Amazingly, unnoticed money saving opportunities are often staring everyone in the face.

Concept: For example: a small 0.5 GPM purge can flow nearly 263,000 gallons in a year. Small air leaks and leaking steam traps can waste thousands of dollars a year and create an undue burden on utility systems. Fixing these mundane problems can be a cheap and easy source of significant savings.

Details: Plants inevitably deteriorate as they age. Because problems develop slowly and soon become familiar, plant personnel often fail to notice slowly developing air line leaks, sticking steam traps, and minor valve leaks throughout the facility. However, these problems can needlessly waste tens of thousands of dollars a year. Consider a steam trap that continuously leaks a tiny flow of 20 #/hr of 600 psi steam. This trap will leak 20 #/hr × 24 hr/day × 365 day/yr = 175,000 pounds of 600 psi steam a year! Figure the cost of that steam plus the cost of throwing away 175,000 pounds of hot, chemically treated boiler feed water and that single, very small steam leak will waste about $2200.00 a year.

Any continuous waste stream can cost a company a huge amount of money and be a source of potential savings. Look for such things as continuous water purges to pump seals that do not have to run all the time, especially if those seal flows contribute to the overall waste stream of the plant. If the plant is considering an air compressor size increase, it may be possible to avoid having to replace the existing unit by simply working through the plant and eliminating air leaks. Any steam leak is an excellent source of savings because the leak is not only a waste of steam but also a waste of condensate along with the heat energy in that condensate. Ultrasonic leak sensors can quickly locate many steam and air leaks even at a distance.

Watch-Outs: Plant personnel will occasionally make an equipment change to address a problem in the middle of the night with no thought to the long term impact of the modification. Here are some "late night" impromptu fixes that seem great in the short term but can have long term implications for unnecessary waste if left in place:

- A condensate pump fails in the middle of the night so Operations just opens a valve and lets the condensate drain out of a reboiler to the deck in order to keep the exchanger in service.
- A pump bearing is overheating so the operator throws a water hose on the pump to keep the bearing cool.
- Heat trace fails on a line so Maintenance winds some steam tubing around the line and crimps the end to let it blow and keep hot.

Any of these changes solves the problem at hand but if left in place they may ultimately waste thousands of dollars. If "temporary" fixes are implemented by Maintenance and/or Production that involve such things as continuous purges, be sure to follow up and make certain they ARE temporary.

Exceptions: Do not immediately assume that every purge or continuous flow can be eliminated. There may be a very good reason for that flow and removing it could have detrimental consequences to the process. Always ask the operators and/or production supervision about the flow to find out why it exists.

Insight: Many steam trap vendors will offer to do a free or very low cost "steam trap survey" to locate leaking steam traps in your facility. Obviously they are looking to sell your company new steam traps, but the leaks they find and eliminate will more than offset the cost of a couple of replacement traps.

If you have any compressors in your plant, check the status of any recycle/blow off/anti-surge valves. Most compressors require some type of an anti-surge valve on the discharge of the compressor which will automatically open if the compressor flow goes too low. Air compressors will usually just vent the discharge to atmosphere. Vapor (non-air) compressors will usually return the discharge back to the suction in order to keep the flow through the compressor above the minimum flow for the machine. If the anti-surge controller is poorly tuned or if the valve leaks, a tremendous amount of energy can be wasted continuously. If designed correctly, these valves should never be open in normal operation. If the anti-surge valves on your compressors are routinely allowing vapor (or air) to pass through them, investigate to see if changes cannot be implemented to close them.

Rule of Thumb: Any continuous waste stream is an excellent candidate for a money saving opportunity. Always keep an eye out for them.

Tip #44: Foster Relationships with Honest Vendors

Hunter Vegas (Career)

Automation engineers spend a lot of money. Instruments and control systems are expensive and in many cases the project manager is entirely dependent on his or her automation engineers to select the best equipment for the job. The best automation engineers make every effort to keep abreast of the latest equipment offerings, but they can never know everything so they often turn to the vendors for expertise.

The engineer/vendor relationship can be close in all forms of project engineering, but a close relationship is particularly important in automation due to the high rate of product development and improvement. While such a relationship can be beneficial to both parties, it can also be a source of serious problems if all dealings are not kept honest, open, and above reproach. The large amount of money involved can be a huge temptation for ethical misbehavior.

My first mentor, Henry Hecht, had a vendor offer him a new boat if Henry would grant the vendor the sole source contract for the control valves on a large expansion project. Henry promptly threw him out of his office and banned the vendor from the plant. Unfortunately, engineers sometimes fall victim to such offers and often pay a high price for their misdeeds.

In comparison to Henry's experience, I have had vendors who were true partners. I could count on them to provide technically valuable and unbiased advice, fair pricing, and excellent service and support after the sale. A vendor representative once literally got up at 2 AM, tossed a valve into his car, and drove an hour and a half to deliver it to my plant because we were facing an emergency and were on the verge of an unplanned and extremely costly shutdown.

Concept: Throughout your career you will deal with a multitude of vendors. Some are quintessential used car salesmen, promising anything to clinch the sale. Others are straight up, truthfully revealing the strengths and weaknesses of their products and offering fair pricing. When you encounter such a vendor, it is wise to foster the relationship. There will be times in your career when you will need an instrument replaced or a spare part on an emergency basis and those vendor relationships will allow you to get equipment and support when nobody else can.

Details: As mentioned in the opening, vendor relationships are crucial to the success of automation engineers. Despite herculean effort on your part, it will be nearly impossible to keep up with the rapid pace of product development, and vendors are a primary source of technical information. Beyond that, technical advances have made instrumentation critical for plant operation and an unexpected failure can cripple or even stop production until the instrument is replaced or repaired. In such late night emergencies (no instrument EVER fails during the day shift!) a strong vendor relationship can make the difference between starting up in an hour and starting up in a few days.

Of course, the vendor is dependent on the engineer for sales and every order increases his bottom line. In the best case scenario, both parties value each other and both benefit. The engineer gets sound technical advice, solid service and support, and a good product at a fair price. The vendor gets significant repeat business and a decent profit. Unfortunately, it takes willingness and effort on the part of both parties to get there. Each must be honest and open and both must believe that the relationship benefits each of them equally. If the engineer (or his company) is constantly squeezing the vendor for lower prices, then the vendor will start playing pricing games to get his profit back or find other clients who are easier to manage. Similarly, if the vendor is concerned only about short term profit then he will say anything to get the order and not be concerned about service or support in the long run.

Watch-Outs: Beware of any vendor who bad-mouths his competition and spends more time disparaging the competitor's products than selling his own. If a sales person has a good product to sell, he or she will rarely even mention the competition.

Also beware of getting too comfortable with a vendor. If a vendor relationship is too good, there is an incentive to enter a sole source arrangement. While this arrangement may offer good pricing and service in the short run, the price breaks usually deteriorate over time, as do service and support. Try to have two strong vendors for each product. This will generate enough revenue to sustain both while keeping competition healthy.

Regardless of the relationship, keep project bidding above-board and transparent. Make certain each vendor has the same information and the bids are evaluated simultaneously and impartially. There can be a temptation to "help out" a favored vendor by revealing other bids or letting them "re-bid." Avoid this behavior at all costs.

Exceptions: None. Vendors are only the enemy when you treat them as such.

Insight: Over the course of years, many sales people and engineers ultimately become good friends. There is nothing wrong with that conceptually, but it important to maintain (and be seen to maintain) a solid line between work and personal time. Leave discussions of bids, projects, etc. at the office.

Rule of Thumb: An engineer is as dependent on the vendors as they are dependent upon the engineer. Foster relationships with honest vendors and keep all bidding strictly impartial. To be a successful engineer or project manager, you have to develop relationships with vendors just like you do with customers, co-workers and subordinates.

Tip #45: Learn to Estimate Jobs

Hunter Vegas (Career)

When I started my career I had no idea what anything cost. I would be asked to estimate a small job and I might as well have used dice, darts, or a Magic 8-Ball to divine a number. In an effort to learn I started logging the price of everything I bought – control valves, pressure transmitters, terminal blocks, junction boxes, etc. Eventually I reached a point where I could accurately judge the cost of equipment without consulting a vendor for prices.

Estimating labor was much more difficult. First I had to learn exactly how much labor was required for various technical and engineering tasks and then I had to learn the various "fudge factors" to apply to account for who was actually doing the work. If the normal on-site contractor was doing the installation they were relatively efficient and consistent, and easy to estimate. On the other hand, some sites had maintenance crews who occasionally would get assigned odd jobs if they were not busy. Suddenly, the hours would start mounting and my estimate might be off by a factor of two or even three.

Over the years I have developed elaborate spreadsheets that allow me to estimate projects from simple to dizzyingly complex. Keeping those spreadsheets accurate is an ongoing effort, requiring constant refinement as equipment prices change and labor rates shift, and I take the time after each project to compare my estimate and the actual cost to see where I was off.

Concept: Estimating jobs is an invaluable skill that every automation engineer should master as soon as possible. The best engineers can size up a job fairly quickly, generating reasonably accurate estimates of cost, schedule, and labor with limited information.

Details: All new engineers are awful at estimating. No college course covers this subject and estimating is closer to an art form than an engineering skill. The ability to accurately estimate jobs is critical to the success of a project engineer, but all engineers need the ability to quickly produce "down and dirty" estimates based on incomplete and imperfect information as well as generate accurate and detailed estimates when more information comes available. The first skill is necessary for quickly evaluating the feasibility of a project and determining if the payback is favorable. A preliminary cost estimate/ project savings comparison can enable you to easily pick project "gems" from a rubble of ideas and not waste time pursuing projects that have no justification. The second skill is required to accurately determine exactly what a project will cost and obtain capital funding once a project is approved.

An engineer learns estimating by doing it. As a new engineer you should try to estimate every job you can, even if you are not responsible for project budgeting/estimating. When you are given a project, take a moment to try to determine the amount of material, technician hours, CAD hours, and engineering hours that will be required before looking at the budget. (This is a good habit regardless as it is much better to know a project is woefully under-budgeted BEFORE beginning it!) After the project, follow up and compare the actual material and labor figures with your initial estimate. Note the differences, learn from mistakes, and try again on the next job. Estimating truly is a skill that must be learned by experience.

Watch-Outs: Risk = Money. When you are estimating a job, it is important to realize that unknowns pose risk and risk equals money. If there are many unknowns in a job (poor specifications, vague requirements, unknown technology, etc.) then the estimate's contingency percentage must be increased. If such a project estimate is too high for a client, work to identify and resolve the unknowns so that a more accurate (and hopefully lower!) cost can be developed.

Exceptions: Some companies have a capital project estimating group (or dedicated estimators) that do all of the estimating. These estimators usually have the task down to high art, using an array of software packages, personal spreadsheets, and experience to generate amazingly accurate estimates for multimillion dollar projects with essentially no information at all. If such a group exists in your company, you may not have the opportunity to estimate work. Even so, try to learn it anyway. Estimating may not be in your job description today, but it almost certainly will be at some point in your career.

Insight: Everybody estimates differently. Some try to identify every single task and piece of equipment and dutifully record it all in a huge spreadsheet to determine a project cost. Others use a more holistic approach, estimating task groups and materials as single entities. Either method can be accurate if done correctly. Many times several project engineers will estimate a project and despite different approaches they will all come back with figures that are within a fraction of a percent of each other.

When you are reviewing an estimate prepared by someone else it is usually best to independently "run the numbers" and compare the two results. It is extremely difficult to check an estimate line by line unless the estimating style is very similar to your own.

Rule of Thumb: Estimating is an invaluable skill for any engineer, and the only practical way to learn it is do it repeatedly. Estimate each job before starting it and always follow up and compare the final cost figures against the initial estimate.

> **"The best engineers can size up a job fairly quickly."**

Tip #46: Success Breeds Success and Mediocrity Is Never Acceptable

Hunter Vegas (Career)

Study after study has shown that human perception quickly turns into reality. If a class is told they are troublemakers and slow learners, then over time that is exactly what they will become. However, if that same class is treated as an elite group where expectations are high and good grades are simply a given, then that class will usually rise to the occasion and deliver a much higher level of performance.

As a father of two boys I have coached some 30 seasons of soccer teams aged four through fourteen. From the outset I quickly realized that the team's perception of itself ultimately determined how they would perform. Most new coaches treated the four or five year olds AS four or five year olds and had very low expectations of their abilities. I treated my team as true soccer players, constantly pushing their skills well beyond their years. Imagine a team of four year olds with set plays, moving down the field passing the ball, and everyone knowing they had a job they were expected to perform. We had great fun, and usually kept playing well past practice time because the kids did not want to stop. All the while the kids were learning the game, learning how to play smart, and learning how to win. They were also learning that everyone played a part – even the weaker players had special tasks that were critical to the success of the team. When we walked onto the field the kids KNEW that they could win – and they almost always did. Over 30 seasons the teams chalked up a 90+% win rate. They also learned to win – and lose – with style. They were never allowed to bad-mouth or trash talk the other team, "in your face" goal celebrations were forbidden, and at the end of the game we always shook the other team's hands, looked them in the eye and said "good game" and meant it – regardless of who won. Many of those same kids went on to play the highest levels of soccer or excel in any number of other athletic endeavors.

Managing a successful project team is really no different.

Concept: Never accept a mediocre effort as "good enough." Expect and demand the highest quality as a given and consistently demonstrate that by setting a good example as a leader. Teams will usually mirror the work ethic of the leader. If you slack off – so will they.

Details: A successful team usually consists of the following:

- A strong leader – The leader must be good at communication, reasonably familiar with the same skill sets as the team, able to quickly gather information and make a decision, and able to admit when he or she is wrong. Most importantly, the leader must earn the trust and respect of the team. If the leader does not have the team's trust and respect, the team will almost certainly languish and will ultimately fail.
- Senior team members – Most teams have at least one seasoned team member who is technically skilled in his or her role and comfortable with the leader. Like the team

leader, these members must also be willing to lead by example and to help mentor and develop the newer team members.

- Team members – The remaining members must be hard working and must either know their roles or be willing to learn them quickly. All team members should take every opportunity to learn new skills and/or technology whenever possible.
- A winning mindset – The team members must perceive themselves as winners and believe that the project's ultimate success is essentially a given. Through his or her words and actions, the leader should make certain that the team understands that failure is simply not an option and that the team must do whatever it takes to complete the project on time, on budget, and with a satisfied client. A team that believes it will be successful usually IS successful, and after a chain of successful projects, that team is usually considered the "premier team" of the office/plant. The members are ultimately recognized (and paid) accordingly.

Watch-Outs: As a leader you must remember that it is too easy to delegate the work to the team members and let them carry the load. There is often also a tendency to favor some team members over others and treat them differently. This erodes trust and respect and can quickly poison a team's dynamic.

The best leaders work as hard as (or harder than) the team, and they are not afraid to jump in and pick up whatever task needs to be done to get the job finished.

Exceptions: Despite all of the politically correct propaganda, everybody is NOT a winner and the skill sets and capabilities of individuals are different. When creating a team, pick those team members that offer the right skills, the necessary technical capabilities, and the ability to work together. Do not be concerned with other people's perception of what the team mix should be. Some will think the team should consist of members with more seniority, others will say that gender and cultural diversity are most important, and still others will seek to exclude certain groups or individuals. The best team will consist of those team members with the right combination of skills, work ethic, and desire to get the job done. This resulting mix will almost never match what others think it should be.

Insight: The leader sets the stage for the team. If he is lazy the team will likely follow his example. If he is divisive, the members will quickly begin bickering amongst themselves. However, if the leader is respectful, talented, and hardworking and picks the right group of co-workers, project success is virtually assured.

Rule of Thumb: Pick your team members wisely, lead by example, and always convey the highest expectations of performance. Your team will deliver.

Tip #47: Grow Your Direct Reports

Hunter Vegas (Career)

I was extremely fortunate as a young engineer. Early in my career I took a position as an "instrument engineer" working under a gentleman named Henry Hecht (whom we met earlier). Henry had 50+ years of experience in instrumentation and controls and the other team member, Clyde Barre, had 45 years of experience. I was a little surprised that Henry and Clyde even hired me because I was totally green and knew essentially nothing about the field. (I think they were looking for a young buck to check out the instruments on the upper floors and on the tops of the 200′ distillation columns but they would never admit to that!) Whatever the reason, they took me on and from the very outset I badgered them with questions. I wanted to understand everything and with infinite patience those two men taught me everything they could. As control technology shifted from "wind powered" pneumatics to digital controls, my electrical and computer background gave me a distinct advantage and some bosses might have felt threatened. Not Henry and Clyde. They considered it their DUTY to teach me and in fact we taught each other. I learned about pneumatics and control theory in general, and I taught them about computers, fiber optics, and the digital world. Ultimately they both retired and I eventually assumed the role of Group Leader, but the lesson of growing your direct reports was not lost on me.

Concept: As a leader/boss, you should do everything possible to develop your direct reports. Broaden their skill sets, give them credit when they perform well, encourage them when they fail, and train them to take over your job. If you are "irreplaceable" then you can never leave the position you are in!

Details: One of the inherent duties of any leadership position is to train your direct reports to take over your job. Even if that never happens, there is nothing but upside to having competent team members who can take over when you are busy or unavailable. On the other hand, nothing can divide and destroy a team quicker than a boss/leader who intentionally "holds people down" to protect his or her current position or serve his own interests. Ironically such behavior may drive the team members to transfer out, get promoted, and eventually become the leader's boss!

There is no downside to teaching your direct reports everything possible, and do not limit your tutelage to technical matters only. Young engineers often need direction on interpersonal skills, client management, and a whole host of other non-engineering subjects. At times it may feel more like parenting then engineering, but the end result will be all good.

Watch-Outs: Be wary of any boss who intentionally withholds information or refuses to take the time to teach or explain something. Working for a suspicious dictator is not enjoyable!

Also beware of a boss who takes credit for the work of his team but fails to credit the members who actually did the work. This is a sure sign of a manager who does NOT have his team's interests at heart.

Exceptions: None.

Insight: For a senior engineer or project manager, watching a fresh out of school, green engineer blossom into a competent and sought after professional is wonderfully fulfilling. It almost IS like parenting, except they rarely ask to borrow your car.

Rule of Thumb: Developing direct reports is a fundamental duty of the boss/leader position. Take the time and make the effort to teach and grow the next generation of engineering talent.

"Working for a suspicious dictator is not enjoyable!"

Tip #48: Treat Everyone with Respect and Do Not Accept a Lack of Respect from Others

Hunter Vegas (Career)

I was born in a very small town in southern Mississippi. From the earliest age, I was taught that every adult was a "ma'am" or a "sir", that you always held open doors or gave up your seat on the bus, and that everyone deserved to be treated with respect and courtesy. Later in life I was occasionally berated by some women who felt they were "too young to be called "ma'am." I apologized and tried to accommodate their wishes, but the fact is that I say "Yes ma'am" to any lady from 2 to 102 years old and after doing it for this long it feels like the right thing to do.

Beyond saying "ma'am" and "sir" where it is appropriate I try to treat co-workers and clients with a high level of professional respect. I certainly do not always agree with everyone and you can be certain they do not always agree with me, but regardless of our differences I try to maintain a high level of decorum and courtesy and I expect the same from them. I do not bully people and I will not allow others to bully me. I have had domineering clients try the bullying tactic early in a relationship and I have immediately called them on it. In almost every case the client apologized and we ultimately worked well together. On the rare occasions when a client demanded the right to treat me or my team with disrespect, I simply walked away from the project.

Concept: Treat people as you would want to be treated. Arrive at meetings on time and be prepared. Show respect to others and expect the same from them. Never allow a client to bully you or your team. It is better drop a client of that sort rather than endure their actions.

Details: Conflict is inevitable in the automation engineering field. There will always be differences of opinion (and some may be strong) concerning project design, money and budgets, realistic schedules, client demands, etc. Do not let this conflict turn into chaos by losing sight of appropriate decorum and professionalism. Arguing is fine and defending a position is certainly warranted if your position is correct, but do not resort to personal attacks as a means to achieve your goal. Such activity may win the battle but it almost always loses the war. When a problem transitions from a technical issue to an emotional issue, the situation becomes much more complex and the ultimate resolution will become infinitely more difficult to achieve. Even when the issues are finally resolved, the emotional scars and resentment can linger for years. People can be extremely slow to forget.

Being respectful of others is important, but you should also expect similar behavior from others. Do not let people routinely stand you up, waste your time, or publicly disrespect you. This is true for co-workers, bosses, and even clients. As mentioned in Tip #44, clients and vendors need each other and the best scenario is a situation where the vendor and client act as a mutually supporting team. (In this case the engineer is the "vendor" and the plant the "client".) Unfortunately some clients will try to establish a dominant position from the outset suggesting that they are in charge and if the vendor does not comply with their demands they will be replaced in an instant. This is almost always a false premise. Unless one is buying a generic commodity, the list of acceptable candidates for a particular piece of equipment or service is limited and in most cases the client has selected a particular vendor that is best suited for a particular project. Therefore, despite what the bullying client might suggest, the vendor is actually in a much stronger position than it may seem. If the client tries to bully the team, the wisest course of action is to immediately call out the client on their behavior and set the precedent for how everyone will be treated going forward. In many cases the client will develop a new respect for your having stood up to him and a long term relationship will result. In some rare cases the client will demand to get their way and at that point an immediate exit is the best strategy. Sending such a problematic client to the competition rids your firm of a long term troublesome client and deposits the same in your competitor's lap.

Watch-Outs: Be particularly wary of clients who will berate you in front of their co-workers and then apologize in private afterwards. If this happens, point out that the behavior is unacceptable and will not be tolerated again. If it is repeated simply walk out.

Exceptions: Occasionally a firm will sign a contract or make a commitment to a client such that walking out is not an option and the precedent of bullying by the client has been established. In such a case the on-site staff is forced to endure whatever pain the client decides to inflict. This is a prime case for the application of Tip #3 – Pain Is Instructive. Have the on-site staff refer all client communications to the manager and/or sales person who set up the contract and/or commitment in the first place. Now the person responsible for creating the situation must deal with the unruly client on a daily basis and must suffer the pain of their contractual decision. Either the situation will improve or the manager/sales person in question will likely never agree to such an arrangement again!

Insight: This concept is particularly applicable to union/management relations. After years of contentious negotiations, some unions and/or managers develop strong-arm tactics to get their way or grandstand in front of the other union members. Like all union/management issues, precedent is the key. Once the precedent of treating each other poorly has been set, changing it can be very difficult.

If you are promoted into a management position that involves union relations you will have one opportunity to set the precedent for future meeting behavior. (Unfortunately you cannot reset the precedent for contract rules, etc. – those have been set by prior management decisions and you must abide by them.) If the Union Committee comes storming into your office screaming and shouting, promptly throw them out of your office and tell them they can return tomorrow morning if they are prepared to discuss their issues calmly. Start the next meeting with an explanation that despite what previous managers might have allowed, screaming, shouting, and other disrespectful behavior will never occur in any future meeting. They are welcome to disagree and discuss the issues, but they are not welcome to act like five year olds. If they DO act like five year olds, they will be dismissed from the room....then DO IT! Even historically stormy relationships can be calmed if both sides treat each other with decorum and respect, and are consistent, honest, and fair.

Rule of Thumb: In any relationship, set a precedent of mutual respect and professionalism from the very beginning. Once the precedent for a relationship has been set, it is difficult to change.

Tip #49: Working For a Boss Whom You Do Not Respect Is Difficult

Hunter Vegas (Career)

This tip is really an extension of the two previous tips and it discusses what can happen when those two tips are not followed.

I have had a number of bosses during my career. Most were superb, providing professional and technical guidance and support when I needed it and giving me the reins and responsibility when they thought I was ready. However, two of those bosses were truly awful.

In both cases they became my boss through company personnel moves over which I had no control. Both individuals were manipulative and technically inept, but incapable of being wrong even though they were often wrong on a daily basis. In each case I stayed in the job about a year, thinking that I should give the situation and the boss at least twelve months before I considered any moves. Unfortunately, after a year each boss had proven to be even worse than I had anticipated and I left those companies, never to return.

Back in Tip #33, I mentioned that high pay is not reason enough to stay in a bad job. Well, nothing can make you dislike a job more quickly than working for a boss you despise.

Concept: When interviewing for a position, interview your perspective boss as hard as he (or she) is interviewing you. The pay may be great and the position fantastic, but a bad boss can negate all of that.

Details: As should be clear by now, the boss sets the tone for the whole team dynamic. His level of effort, his desire to train and develop his direct reports, and his management style will in large part determine the working atmosphere along with the ultimate success of the team. Great co-workers are important and wonderful to have, but the boss's personality will more likely determine whether you actually like your job.

Because the boss's management style is so critical, it is imperative that you vet him/her thoroughly before accepting a job. Go out to lunch with the boss and other team members and see how they interact. Invite the boss to dinner with his/her spouse and watch how they treat their spouse and the wait staff. Is the team comfortable with the boss and does the conversation flow easily, or are they quiet? Does the boss mistreat his/her spouse or talk down to the restaurant staff? Is the boss decisive? Can he/she admit fault? Ask team members if there are any individuals who left the group. Is so, track them down and find out why they left. Take the time and make the effort to find out all you can about the boss because your level of job satisfaction will in no small measure be determined by this one person.

Watch-Outs: If you are touring the plant or office and the boss lies to other workers or team members, talks down to his direct reports or treats lower level company staff with disdain, moving on to other possibilities is probably the best course of action. The boss may be brilliantly successful and knowledgeable but working for a tyrant can be unbearable.

Exceptions: Some companies transfer "Hi-Po"s (High Potential Management Employees) from job to job every 12 months. If you get one assigned to your group and they are awful, it may just be easier to bite your tongue and wait them out.

Insight: A good boss can even make staying in a bad job worthwhile. The company may be struggling or the corporate politics ugly, but a good boss will somehow find a way to keep the team together and shield them from the fray. However if things get too bad, a good boss is one who encourages his direct reports to develop and grow their careers, even if that means leaving the group or the company.

Rule of Thumb: A bad boss should be a deal breaker of any job consideration. Interview the boss thoroughly and carefully before you accept any job.

"A good boss can even make staying in a bad job worthwhile."

Tip #50: Ask for Feedback

Hunter Vegas (Career)

And so I have finally reached my last tip. Trust me when I say that this has NOT been an easy task. Distilling a career of engineering experience into 50 one or two page tips is tough!

For my last tip I chose this subject because I believe an individual should never stop learning and improving ... and you cannot improve unless you know what you are doing wrong. The value of following this tip may seem obvious, especially to a young engineer, but it is not so for an older, more experienced professional. As engineers rise up the technical ladder and gain more prestige, asking for feedback often becomes increasingly difficult. Ego and pride get in the way and it becomes difficult to admit fault or error. As soon as that mindset creeps in, further improvement becomes impossible and decline is inevitable. The fact is that nobody is perfect and regardless of how high in his or her career an engineer might be, constant adjustments are necessary.

Concept: After wiring a panel that you designed or working with a graphic that you created, ask for feedback from the technicians or operators. At the end of a large job, pull the team together and review the project to see what went well and what did not. Always seek to improve everything you do.

Details: Any day that passes without learning something is a wasted day. Continuous growth and improvement should be your goal regardless of your level of experience. The only way to accomplish that is to ask for feedback.

Expand beyond the usual list of client feedback forms or follow-up phone calls that most companies employ. What did the operators like/dislike about the system? Has Maintenance had any problems with a piece of equipment since it was installed? Do the technicians have any suggestions for improving a panel design to make it easier to fabricate or troubleshoot? Does the electrical contractor have any ideas for improving the design/installation documentation for the next project? All of these people can provide valuable ideas for improvement and you are probably the first person to actually ask their opinion. Ask the question, and then LISTEN to what they have to say.

Watch-Outs: Do not get defensive. Most people feel the need to immediately defend their work against criticism. No matter how well founded the criticism, resist this urge. If the criticism seems unwarranted, it is acceptable to ask further questions to determine exactly why the person feels as they do, but it is not acceptable to deny or disparage that person's point of view. Getting defensive will ensure that you will never receive any feedback from that person again.

Exceptions: If you are a leader, asking your direct reports for feedback can be problematic. They are understandably motivated to say only positive things for fear (however unfounded) that any negative comments could be held against them. It may be possible to establish a strong relationship with a senior team member and let him or her solicit general feedback from the team without specifically attributing a comment to a particular member. Sometimes other senior managers or project leaders can gather feedback from the team through other venues (such as performance reviews, etc.) or during casual conversation.

Insight: Possibly the most illuminating feedback will come from close friends or colleagues. These people, above all others, are in a position to straighten you out when you need it. When this happens fight back the defensive response, listen to what they have to say, and take it to heart.

Rule of Thumb: Continuous improvement requires the continuous solicitation of feedback. Once that feedback is received, ACCEPT it and act on it.

> **"Ego and pride get in the way and it becomes difficult to admit fault or error."**

Tip #51: Seek Conversations with Knowledgeable People

Greg McMillan (Philosophical)

I got a degree in Engineering Physics, a now-defunct program at Kansas University that was created for nuclear submarine captains. I did double duty, taking the required courses for a physics degree plus 32 hours of engineering electives. After graduation I got job offers from aerospace, chemical, and communication companies. I chose the chemical company because the campus was impressive, and, as an instrument engineer, I would have my own cubicle. I had no idea what an instrument engineer did. I had never even heard of the profession until my interview. I didn't expect to use my courses in Quantum Mechanics, Statistical Physics, and Astrophysics (although I would later see the Uncertainty Principle, Spin, and Relativity applied to corporate decisions). I was unprepared for the fact that nothing I had learned in getting my degree would be used on the job except for a few concepts from courses in heat transfer and fluid mechanics and an electronics lab for test instrumentation.

My subsequent career in process automation turned out to be an exceptionally fortunate choice, leading to more opportunities than I could have ever imagined, mostly because of the people I could talk to. My mentor from 1969 to 1971 was an instrument technician who was responsible for the instrument design in a new process that used three renovated production units and was scheduled to start up within two years. He did not have a degree but had 20+ years of experience and the friendliest and kindest attitude I had ever encountered. His personality and experience were only exceeded by that of our supervisor. I was immediately sent to the field in electrical and instrument (E&I) construction to make sure the E&I systems were properly calibrated, installed, and commissioned. Like me, the union pipefitters and electricians assigned had no experience with instruments, valves, control rooms, and motor control centers. We learned together. The start-up was difficult at best due to unknown process problems and rusty on-off valves posing as throttling valves, but, in the end, all of the units came online successfully. I learned 10 years of stuff in two years. I went on to other construction and design assignments, taking advantage of the knowledge of Stan Weiner, future co-author of ISA books and the *Control* magazine Control Talk column. I then got an invitation to move to Engineering Technology after doing an extra-curricular dynamic simulation of compressor control. What an incredible opportunity! I gained access to the world's best in modeling and control, which led me to finding great articles and books and seeking the experts at the suppliers. I asked a great many questions, but I never had a person refuse to talk to me no matter how important he/she was or how unimportant I was.

To this day, I enjoy great discussions with automation professionals by virtue of my *Control Talk* column. There is no real interview. We just talk for a couple of hours, and the result is a column that captures essential practical knowledge that is often not documented. I met Hunter Vegas this way. Our three-part Control Talk series and now this book are great examples of the value of connecting diverse expertise.

Concept: Success as an automation professional requires knowledge of many areas including chemistry, DCS configuration, process dynamics, equipment, instruments, material and energy balances, prime movers, physical properties, piping, unit operations, valves, and variable frequency drives. You are a generalist and often dependent on the expertise of others, expertise that is gained on the job and is largely undocumented. Communication, interpersonal skills, and an open mind are more important today because no one person can have all of the knowledge. Successful implementation requires being a team player (Tip #57). Good conversations can be mutually beneficial for opening doors and broadening horizons.

Details: As a practical application of seeking conversations with knowledgeable people, do your homework by talking with associates, searching the Internet, and checking out handbook sections on the subject, then stop by the offices of process, mechanical, E&I, and configuration engineers in your plant or corporate offices with an intelligent question. Armed with knowledge, call the corporate office of a supplier and ask for technical support. See what the supplier knows about the product in terms of explaining what you learned. Provide follow-up results to anyone who gave you advice. Engineers love to solve problems. Take advantage of this desire and talent.

Go to the instrument shop and get to know the technicians. See what they are working on and gradually ask more and more questions about what works and what doesn't. When in a plant, spend each morning in a control room to see what the operators are doing. Become aware of problems operators are having with loops. Review with the operators trends for oscillations and trajectories. See the "Checklist for Loop Analysis by Trend Charts" in Appendix C for more details.

Watch-Outs: Experts can be wrong (humans are not perfect). The more prominent and specialized the expert, the bigger the ego and the greater the probability that some factor outside his knowledge may be in play as you seek his advice on a problem. If what you are hearing is clearly wrong, provide polite feedback but be prepared for denial of what you observed and the reiteration of the original advice. Do not try to change the expert's mind. Just move on and do not hesitate to seek advice from the same expert for the next problem. Do not expect expert systems to capture this expertise. Expert systems were the rage in the late 1980s. Nearly all expert systems in the process industry went out of service often as soon as the expert left the control room. Incessant chatter and a few false alarms ended up getting these systems turned off sooner than later. Don't be an expert system in the control room. Don't talk too much, criticize, or raise undue concern.

Exceptions: Do not ask Maintenance unnecessary questions during a turnaround or distract Operations during a shutdown, a start-up, or a flurry of alarms.

Insight: Engineers, technicians, and operators will tell you what they know if your questions are intelligent, informed, and non-confrontational.

Rule of Thumb: Start a conversation with everyone responsible for the successful implementation, maintenance, and operation of your designs with a friendly, open, humble, and patient approach.

Tip #52: Be a Good Listener

Greg McMillan (Philosophical)

Engineers love to solve problems. If your questions are intelligent and you appear conscientious and dutiful in understanding and using the information for a solution, the office door will generally be open for you. A person can tell if you are listening. I use a notebook to capture key ideas and details. Not only does this help in retrieval of the knowledge offered over time, the act of taking notes impresses upon the expert that what he/she is saying is important and will be used. I enter with an open mind and an eagerness to acquire knowledge that is often undocumented and even unspoken.

The discussion should not be driven in a preconceived direction. The conversation should be free to take new directions as insights are gained. Intuitive reasoning as a complement to analytical reasoning is the key to success. Of course, you need to make sure the original problem is addressed but there is a definite advantage in allowing the conversation to pursue unforeseen opportunities. When there is a pause, be quiet and wait; you need to let the expert have ample time to restart.

I have extensively used this simple approach in my Control Talk column for the last 10 years to capture and share the expertise of people whom I know have something to offer. The process is totally different than the interview method traditionally used. There is no list of questions, no agenda, and no sales pitch. For Control Talk, the conversation goes wherever it naturally flows. I usually end up with four pages of notes for a one hour conversation. The whole process is casual, open, and enjoyable. If you become a really good listener and promoter of conversation, you might become the next Oprah (OK, maybe not).

If some piece of advice proves to be wrong, you must inform the expert but don't forget to mention what you learned in the advice that helped you to solve the problem. Make sure the expert remains part of the solution. This approach avoids making the expert defensive, shutting off future discussions. I have found that the greater the expertise, the greater the ego and the need to be right.

Learning from operators and maintenance technicians is extremely important because they are on the front lines. You generally need to establish a personal relationship before such individuals will open up. This requires time and patience. The best thing you can do is hang out in the control room and instrument shop and observe without interfering. Becoming an insider rather than an outsider is essential. The worst thing you can do is to offer your opinions unasked. You may need to be proactive in asking questions because the arenas of knowledge of engineers and technicians/operators are so different. You also need to seek observations more than conclusions because knowledge of causal first principles and automation systems is rare in operations and maintenance personnel. War stories often rule. Getting the observations of different shifts also helps with perspective, and further discussions with operations supervision and process technology support are important to sort out fact from fiction.

Concept: Egos and today's fast pace of life tend to encourage sound bites, superficial understanding, and the reinforcement of established practices. When seeking the knowledge of others you need to "chill," build relationships, truly listen, record what you have learned, and share results in a positive, constructive manner.

Details: Develop personal relationships with knowledgeable people in research and engineering, at suppliers, and in control rooms and instrument shops. Start a conversation to solve your problem. First choice is to see them in person, second choice is a phone call, and last choice is email correspondence. Ask an initiating intelligent question. Take time to listen and let the conversation go where it needs to go. Take notes. Share results and be generous in acknowledging the contribution of everyone involved. "Always asking why," emphasized in Tip#1, is the essence of being an engineer.

Watch-Outs: Specialists may have a narrow view and may not realize when their knowledge is on the borderline of another expertise. Preconceptions may rule. There is also a "not invented here" obstacle on both a personal and an organizational level. Be aware that some experts will not be able to recognize or admit that their advice was wrong. If the dialog becomes negative, destructive, or prejudicial, quickly direct the conversation to a different topic. If the statements are clearly false, correct the person and state the truth clearly (but diplomatically). If recommendations are found to cause adverse results, emphasize the detrimental consequences to the point where the mistakes are not repeated.

Exceptions: While writing notes may be friendlier, more casual, and less intimidating, the future is digital. Voice recognition software can work well for capturing ideas if you can interject your thoughts by supplemental typing. The recording device must be inconspicuous and your typing not a distraction.

Insight: The knowledge of individuals can be openly shared if the environment is inquisitive and friendly and egos take a back seat to truth and problem solving.

Rule of Thumb: Develop personal relationships and engage in meaningful dialog to gain technical and first principle knowledge from Research & Development (R&D) and Design, and what works and doesn't work from Operations and Maintenance.

> **"I have found that the greater the expertise, the greater the ego and the need to be right."**

Tip #53: Find and Read Technical Articles and Books[1,2]

Greg McMillan (Career)

I have about 100 binders of articles and papers and 100 books that were gathered between 1974 and 2004. I acquired about a half of the articles and papers from the late Vernon Trevathan along with his expertise in compressor control, pH control, and tuning and control loop performance. I haven't added much since 2004 except for some new books because everything published today is on the Internet. Right now almost half of the binders and books are sitting in our dining room since I am working from home. So far people eating dinner don't seem too interested in reading my collection even though I consider them a treasure of knowledge. These guests don't know what they are missing. Maybe automation professionals today don't know what they are missing. Unfortunately, I can't invite them all over for dinner. Hmm, could a big Texas BBQ be the means, or would we all just get happy from Lone Star beer? Either way, the idea is worth a try. Look for the announcement on ISA Interchange.

I got a scare this past summer due to wildfires burning down whole towns within a couple of miles of my spread (Texas talk). I was told you have five minutes to leave once the police or fire officials knock on your door. I gathered an overnight bag, my senior citizen medicines, and a dozen out-of-print books and booklets that have technical information not found elsewhere. The booklets had test results and equations for pH and ORP electrodes from the inventor and founder of the company for the first accurate and reliable steam sterilizable electrodes. The original company has been bought several times and the new owners don't know these booklets exist. I found a similar problem with Shinskey's books and papers. I need to scan in my collection or at least put them in a spreadsheet. I have scanned in key papers on material and energy balances for extruders for my model being developed in MiMiC. I also scanned in key research papers by Bill Luyben on valve position control, recycle system stability, and the effect of exothermic reactor temperature lags on the window of allowable controller gains.

The best books are largely out of print. Publishers making decisions as to which books will stay in print only look at sales trends. When we didn't hire many automation professionals in the 1980s and 1990s, sales went down. There was no concept of expert or knowledge. How else can you explain why most of Shinskey's books are out of print? A company (UMI - ProQuest) microfilmed books in the late 1990s and early 2000s, where permission was given back to the author but the results were hit or miss. Fortunately, I purchased from UMI the book *The Determination of pH,* which has a hundred times the content of any other book on the physical principles for pH before the reprint company went out of business. As far as I know the UMI microfilm files were lost. My most important books suffered this fate. Fortunately, Momentum Press is putting books like mine back in print.

Concept: Most of the major discoveries in connection with control valves, control strategies, model predictive control (MPC), PID control, and process measurements were made prior to 2004, except for some "smart" features and wireless communications. Publications (other than handbooks) that describe these developments are largely out of print. You can buy used books, but after this book is published the cost of the best may skyrocket. You can get libraries to get reprints of articles and papers for a fee but you need to know the publication details. I see a need to put a list of over a 1000 articles and papers I have collected online as the best of the best.

Details: Find everything you can that was written by Karl Åström, Béla Lipták, Bill Luyben, Greg Shinskey, and Cecil Smith. Other exceptional books (if you want to dig deep into process dynamics) are Peter Harriott's *Process Control* and Roger Frank's *Modeling and Simulation in Chemical Engineering*. If you do not have a Chemical Engineering degree, I suggest Max Peter's *Elementary Chemical Engineering*. You should also have *Perry's Chemical Engineers Handbook*. If you are in the biopharmaceutical industry, I suggest James Bailey's *Biochemical Engineering Fundamentals*. See Appendix B for a list of books that have essential information not being published today.

Watch-Outs: Be wary of oversimplifications and glorifications where selling is the overriding motivation. Hunter's Tip #2 is right to the point with this concern. Unfortunately, the authors listed above are gone or will soon be gone, and, sadly, most professionals today don't know these names. The articles being published today are mostly written by marketing and sales to showcase their products and by academics to showcase their research, without plant experience or an understanding of the knowledge of the authors mentioned above.

Exceptions: For the most part, new publications on pneumatic instruments and controllers are no longer relevant. Beware of any publications that show pneumatic signal lines or differential head flowmeters downstream of control valves because both mistakes indicate a lack of practical experience.

Insight: The automation profession has matured, but more than 80% of the process control opportunities afforded by technological advances are not realized due to a lack of practical knowledge by all types of automation professionals.

Rule of Thumb: Find key articles, papers, and books, and read an average of 10 pages a week.

Tip #54: Look for Opportunities to Improve Plant Profitability

Greg McMillan (Career)

Measurements today have achieved a 10× reduction in installed error by improved sensor technology and compensation for process and ambient conditions. Control valves have achieved a 5× reduction in stiction and backlash by improved internal flow element design and digital positioner technology. Analyzer technology is now smarter and more easily integrated and maintained. PID features such as external-reset limit and adaptive tuning have dramatically increased the ability of the plant or engineer to achieve a broad range of process objectives. Distributed Control Systems (DCS) incorporate standardized function blocks that facilitate creativity. Books by Liptak, Shinskey, and others have documented a broad spectrum of control strategies that may be applied to increase process capacity and efficiency by a simple DCS configuration change.

In the field of automation today we have a convergence of the need to show a positive impact on the company's bottom line and the technologies that allow us to pursue new opportunities for doing that. What we are missing is the people with the time and experience and the knowledge of what is possible. This book hopes to address what is needed to spark imaginations.

When budgets and people were rapidly disappearing in the 1990s, the chemical company I worked for started a process control improvement (PCI) group of six modeling and control specialists that was fully funded to find opportunities. The PCI program was created by the late Vernon Trevathan, developer of the ISA program for Certification of Automation Professionals (CAP). Vernon promised and delivered a 4% reduction in cost of goods in the plants. The fact that Vernon successfully proposed and accomplished this goal shows that the identification and achievement of benefits is a key to the advancement of the individual and the profession.

Vernon started the PCI program with benchmarking studies, where we visited the best companies with the best technologies in terms of process control practices and achievements. We found a diversity of technologies and achievements, as documented in the report I wrote for the Chemical Manufacturers Association (CMA). My CMA report and Engineering Technology presentation on the opportunity assessment methodology are in the June 2012 Control Talk column "The Human Factor."

You can pursue your own PCI initiative by estimating the monetary value of your ideas. Tips #59, 61, and 62 can help. Warning! If you are successful, you may have to put on a suit and dress shoes to make a presentation at a board meeting. You may be forced to eat French pastries and caviar. If this happens, you can do us all a favor if you can get the following quote by you into the *Wall Street Journal*: "Automation systems tell us what is going on in the process and provide the means for us to continuously maintain the process at its best. Automation professionals are in the best position to improve company profitability and shareholder return."

Concept: You need to show the impact on the bottom line to be given the freedom and resources to do more than simply execute control definition projects that were developed by Process Design and Operations. Track down the sources of variability. Get with the process engineers and make simple calculations of the cost or profit opportunities. Use innovative strategies to achieve process improvement objectives and reduce energy, raw material, recycle, and waste treatment costs and increase on-stream time and production rate. Take advantage of new automation system tools and functionality.

Details: Become buddies with a process engineer and confidant of the operators as per Tips #51 and 52. Track down the sources of variability, starting upstream and looking for oscillating loops, on-off control, and manual actions. Use auto-tuning and adaptive tuning tools to improve loop tuning. Use the power of the PID to eliminate on-off control and operator intervention. Eliminate manual control per Tip #69 by automating the best of the operator actions and replacing the scheduling of flows with feedback control (e.g., fed-batch control and automated start-ups). Move on progressively to important unit operations, perhaps starting (for example) with reaction and continuing on with crystallization and separation and ending up with blending and drying. Add override control and valve position control for small optimization opportunities or model predictive control for large optimization opportunities. Finally, take a look at waste treatment system efficiency. Compute online metrics (Tip #61). Use key PID features (Tips #71-72 and 100) to reduce interactions, coordinate loops, eliminate trips, reduce off-spec material, protect equipment, and optimize loops. Use a virtual plant (Tip #99) to do experimentation to find and quantify possible improvements.

Watch-Outs: Large equipment volumes will smooth out upstream oscillations, similar to the attenuating effect of an electronic signal filter on noise. Here the filter time constant is the volume residence time (volume/flow). This filtering action must be considered when estimating the cost of variability downstream. Don't get hung up on actual cost numbers. If necessary, use percentages. The relative improvement and technology are what is important. Operator training and handholding are essential, or else whatever is new will be blamed for anything that goes wrong.

Exceptions: If a feed or utility valve is wide open there is not much you can do to increase production rate if downstream variability is acceptable. Large blend tanks can keep material within spec despite upstream variability. If there are only on-off valves, you are relegated to improving sequences until control valves and PID loops are installed.

Insight: You need to be able to estimate and demonstrate the value of improvements in process efficiency and capacity in order to get a commitment of time and money for process control innovations.

Rule of Thumb: Estimate the benefits from a decrease in variability and an increase production rate and implement solutions where the increase in plant profitability provides sufficient justification.

Tip #55: Use the Best Technology, Even if the Price Tag Is Higher

Greg McMillan (Career)

The biggest mistakes I have seen in the process industry came from trying to cut costs by not using the best technology instrumentation and systems. We tend to forget that measurements are the only windows into the process and controllers and final control elements (e.g., control valves) are the only means of affecting the process. If you can make the recognition of the essential role of automation part of your company's culture, you can write your own ticket (e.g., trip to a plant).

In the days of pneumatic positioners, plants were flying blind so far as knowing what a control valve was actually doing. We didn't realize that positioners out of calibration, poor actuator and positioner sensitivity, improper actuator bench settings, excessive friction, and linkage backlash meant that we were lucky to be within 5% of the desired valve position. A theoretical study in the 1970s said that boosters instead of positioners should be used on fast loops, not realizing the previous non-idealities or the fact that the booster set up a positive feedback loop that would cause a butterfly disc to become unstable. To this day, some companies still try to decide whether or not to spend a thousand dollars on a positioner, jeopardizing the loop and the process.

To more easily justify the cost of a DCS in the 1980s, thermocouple input cards were specified instead of temperature transmitters. The resolution of the wide range input cards of the 1980s and early 1990s was 0.25 degrees, preventing the use of the high controller gains and rate action on temperature loops. In addition, the individual drift of a thermocouple could not be compensated for and the individual error in % of span could not be minimized by narrowing the calibration span per application requirements. Distillation columns and reactors that depended on tight temperature control suffered immensely.

The biggest mistake of yesterday (and still today) is using on-off valves as throttling control valves. Piston actuators, sloppy linkages, and spool type positioners are put on piping valves, making an attractively cheap valve with tight shutoff that meets the piping spec (see Tip #83). Packaged equipment vendors are always trying to be the lowest bidder. Many times any old thing that sort of works is OK. Common shortcuts include field switches and gages instead of transmitters, on-off valves instead of true control valves, orifice flowmeters instead of inline meters (vortex, magnetic, or Coriolis) and bare thermocouples instead of resistance temperature detectors (RTDs) in thermowells.

When onstream time and batches are worth millions, measurement life expectancies are less than a year, and/or drift can be greater than one percent per month, install triplicate measurement devices and use middle signal selection (see Tip #88). We triplicate devices in safety instrumentation systems, but fail to make the same cost-benefit analysis in process control systems.

Concept: Automation professionals need to be proactive and use their knowledge of the performance of the instrumentation and its impact on the process to provide systems that will exceed long-term process requirements. When in doubt, engineers should err on the high side. Future requirements in terms of flexibility and process knowledge for sustainable manufacturing are often underestimated.

Details: Specify instrumentation (measurements and final control elements) with the greatest reliability, lowest life cycle cost, highest turndown, least drift, least uncertainty, and best precision. For valves, the sensitivity component of precision is most important. For measurements, the repeatability component of precision is also important. The life cycle cost includes the cost of hardware, installation, piping, maintenance, and process variability. Choose a Distributed Control System (DCS) that offers the greatest I/O flexibility, most advanced control tools, best configuration capability, and PID features (see Tips #71, 72, 91, and 100). For a ballpark estimate of the cost of process variability, take the square root of the sum of the squares of noise, repeatability, sensitivity, and uncertainty and ask a process engineer what is the impact on the process operation and process analysis. If you don't have time to do the analysis, you can use the following guide: Estimate the life cycle costs of orifice meters as 20× the hardware cost, of vortex meters as 6× the hardware cost, of magmeters as 3× the hardware cost, and Coriolis meters as 2× the hardware cost. For reactors and columns, use RTDs for temperatures < 450 °C. Don't use on-off valves or field switches. Use 3 pH electrodes and middle signal selection (Tip #88).

Watch-Outs: The cost of impulse line winterization and maintenance, meter coefficient uncertainty, measurement noise, and reduced turndown is often not included in the life cycle cost of orifice meters. The cost of straight pipe runs and flow conditioners for orifice meters and vortex meters is often not considered because the cost is in the piping budget. The cost of lower sensitivity, greater drift, and greater risk for electromagnetic interference (EMI) of thermocouples compared to RTDs and the cost of extension wire are often not considered. The life cycle cost of field switches instead of transmitters and software switches often does not take into account the inability to verify their integrity and accuracy, their lower reliability and accuracy, and the loss of a control room signal for process analysis and troubleshooting. The life cycle cost estimates of control valves, dampers, and vanes do not include the process variability and maintenance cost of limit cycles and poor turndown from poor actuator and positioner sensitivity and total valve assembly backlash and stiction. Engineers who have not worked in a plant, who move on from project to project in an instrument design function (no onsite for checkout and start-up) and who do not have a dialog with plant operators, maintenance, and process engineers will mostly just understand hardware costs. These engineers have not seen the impact of instrument selection on operability, maintainability, and profitability.

Exceptions: Coriolis meters can get prohibitively expensive in large pipe sizes. RTDs have a much higher failure rate than thermocouples when vibration occurs, whether due to equipment, flashing, or velocity-induced vibration. Insulation resistance degradation at temperatures above 400 °C can greatly increase RTD errors.

Insight: A focus on saving thousands of dollars up front can result in additional maintenance costs and can have hidden long-term costs of millions of dollars in terms of the loss in process efficiency, flexibility, and capacity.

Rule of Thumb: Do not go with cheaper instrumentation unless you are certain the impact on maintenance and process performance is (and will continue to be) negligible.

Tip #56: Learn New Skills and Explore Technological Advances as an Extracurricular Activity

Greg McMillan (Career)

When I was the lead electrical and instrument (E&I) design engineer for what was then the world's largest acrylonitrile plant, in my free time (when not going to concerts and bars in Harvard Square) I learned how to write a dynamic model of a compressor. Because one huge compressor was supplying the reaction air for the world's largest catalytic reactors, I wanted to know more about surge control. I knew that reactor trips in other plants sometimes caused surge and the shutdown of other reactors. I got tables from the R&D group of the control valve manufacturer that enabled me to estimate the pre-stroke deadtime and stroking time of large actuators for various positioners and boosters. I got the surge curve from the compressor manufacturer, and momentum balance equations from the literature. I can't say it was a great model, but it got the attention of Monsanto Engineering Technology process control group leader Vernon Trevathan. One of Vernon's areas of expertise was compressor surge control. I got a job offer to be a specialist in a department of about 75 of the world's best in process, materials, and mechanical technology and in process modeling and control. If I had not done that compressor model, I would have stayed a design engineer and generalist. My focus would have been specifying and buying instrumentation and configuring systems. At the risk of self-congratulation, I probably would not have become a Senior Fellow, would have not been inducted into the Process Automation Hall of Fame, and would have not gotten the ISA Life Achievement Award.

The point is that engineers have a difficult time getting recognition unless they can do something special. Today, you don't need to write programs but you do need to know how to use programs and how to configure applications. In particular, I recommend learning how to do dynamic modeling. In the building of the model, you will gain a deeper understanding of the process and of control problems and solutions. Running the model for different scenarios and control strategies will enable rapid exploration and quantification of process control improvements (see Tips #62, 98, and 99). New modeling packages with graphical configuration capability make the learning process as an extracurricular activity much more manageable. If you don't have the skills and time, start with generic free online control loop simulators. If you build a virtual plant that successfully relates optimization benefits to golf, you may be invited to play golf with your CEO.

An emerging opportunity that is presently neglected is learning how to mine information from an Asset Management System (AMS) and an Instrument Management System (IMS). These systems are generally underutilized, perhaps because the expertise to get the most of out of these systems falls between groups. The smart instrument and positioner diagnostics and the AMS snap-ons for valve, vibration, ultrasonic, and failure analysis offer incredible capabilities for improving existing automation system performance.

There are many opportunities offered by advanced process control tools. Learning how to use an auto tuner and an adaptive tuner with loop metrics is an essential starting point of any career that seeks to go beyond E&I design and into loop performance.

Concept: Learn how to use dynamic modeling, system analysis, and advanced control tools in your free time. As you demonstrate improvements, management will provide the means for you to continue.

Details: Learn how to get the most out of an asset management system, an instrument maintenance system, and loop tuning and metrics tools. If you want to get more into designing process control strategies and improving the loop dynamics per Tip #70 and optimizing the setpoints of control systems, learn how to use dynamic modeling tools and build a virtual plant. To move beyond single loop optimization, learn how to apply model predictive control (MPC) tools.

Watch-Outs: Most out-of-the-box dynamic models do not include transportation and mixing delays, measurement lags, transmitter damping, measurement noise, analyzer sampling and cycle time, wireless update rates and trigger levels, and control valve backlash and stiction. Consequently, the total loop deadtime is dramatically lower than it would be in an actual plant. Because the maximum controller gain and minimum reset time are inversely and directly (respectively) proportional to the deadtime, controller tuning conclusions are thereby invalidated. If valve pressure drops are not updated by the use of a pressure flow solver and tables of valve flow coefficients cannot be entered, the installed characteristic, and, thus, the process gain and resulting controller gain, will not include the effect of valve gain.

Exceptions: If you have children under the age of 10, you may not have time for personal development unless you can rent a grandparent as offered in the ISA book *The Funnier Side of Retirement for Engineers and People of the Technical Persuasion.*

Insight: Distinguishing yourself as having special skills requires initiative because companies are often not interested in personal development unless there is a clear impact on the bottom line.

Rule of Thumb: Use models to better understand the process and control system relationships and learn how to use tools for instrument system and loop performance analysis to improve the calibration and maintenance of instruments and the tuning of PID controllers.

Tip #57: Be a Team Player

Greg McMillan (Career)

When I started my career, I could pretty much execute the entire automation system design from my desk using company standards, company specification forms, extensive vendor catalogs, and access to a host of company experts. I had the benefit of an eight-week internal training program on instrumentation and controllers. Now the mentors are gone, most of the expertise resides with the supplier, companies do not have time to update and maintain standards, and company technical training programs are extinct. You can't do it all alone. You need to be able to work with people to cover all the bases including Distributed Control System (DCS) configuration, operator interface design, smart instrumentation diagnostics, variable frequency drives, tuning software, asset management systems, data historians, networks, busses, wireless gateways and devices, and electrical distribution systems.

People skills have always been important, but now that the days of working individually are gone you need to be team player. Applying Tips #51 and 52, you need to be able to communicate and make people want to work with you.

You need to be a team player because just as it is in sports, the accomplishments of the team are more important than individual accomplishments. The team includes the supplier of products and the services. Concentrate on your role and see the bigger picture. Do not let your actions disrupt the design and implementation process. The DCS configuration engineer, electrical and instrument engineer, Information Technology (IT) person, control panel graphics designer, piping designer, P&ID drawing designer, mechanical engineer, process engineer, research scientist (particularly for biopharmaceutical processes), operator supervisor, supplier, and plant representative all need to understand and meet each other's needs. You have to allocate your time so that you do not hold up someone else.

For people to be able to depend on you, you must make reasonable, but firm, commitments. The worst thing you can say is "I think I can do it" or "I can do it, if I have time." Be specific as to your own needs in meeting it, and then make a decisive commitment.

Less obvious is the need to be supportive of others and to acknowledge all achievements as a team effort - even if you successfully accomplished a task by yourself. If you implement a great idea, personally thank the team members even when their contribution is not obvious. In memos, reports, articles, papers, and conversations, acknowledge the contributions of others. Avoid trash talk and doing celebration dances in the control room, pointing to trend displays and saying, "Uh huh, that's me."

Concept: Successful implementation of an automation system requires myriad skills and is a team effort. You win or lose as a team. Egos have to take a back seat.

Details: Make sure you are making the most of the skills of team members and team members are making the most of your skills. When something goes well, thank and recognize (in writing) all involved. When things go wrong, don't look for someone to blame or for an excuse, but focus on solving the problem together. Make commitments without vague qualifications. If conditions change, immediately update the commitment without complaint. Use email and websites to keep team members updated on current problems, solutions, decisions, schedule, and cost. Use meetings judiciously. For process control improvement, consider the methodology described in the June 2012 Control Talk column "The Human Factor." Develop a quantitative assessment of the benefits by working with individuals from process technology and operations. Use this opportunity sizing and a process review in a structured concise meeting with process control experts to find opportunities.

Watch-Outs: Don't use email to deal with mistakes or resolve disagreements. Email rapidly escalates misunderstandings and bad feelings for reasons that could be a book in itself. Meetings may take up too much of a team's time. Meetings can get conversations started and get team members on the same page, but detailed guidance and solutions are hard to achieve in a meeting atmosphere. A runaway condition can develop in which meetings take up too much of a team's time, resulting in more meetings on why the team is behind schedule, which in turn takes up more time that is needed for actual work. The result is often that the team makes mistakes and misses completion dates. Studies have found that brainstorming meetings are less effective than they were once thought to be and that successful innovation requires a lot of in-depth personal effort and perseverance. Hans Baumann, founder of the control valve company H.D. Baumann Assoc., Ltd., interviewed in the May 2012 Control Talk column "Control Valve Innovations," was proud that his company had no meeting rooms.

Exceptions: If a team member is too disruptive, the person must be metaphorically sent back to the minor league or development league. If an idea is clearly wrong, the idea must be (diplomatically) rejected.

Insight: Sharing knowledge and recognition, individual effort, and having common goals are all essential for success.

Rule of Thumb: Work together on a one-on-one basis, and seek inspiration and recognition as a team.

Tip #58: Make Yourself More Marketable

Greg McMillan (Career)

I have heard engineers say that they can't go to a technical conference or take a course, or can't do some extracurricular work, without even having asked their manager. Even stranger is when these people go into their manager's office and say what amounts to "I know the company doesn't want to pay for me to do this" or "Doing this will interfere with me getting my work done." If this describes you, your manager may think you do not really want to go, and just want an official reason for not doing something beyond your normal work requirements. The manager may genuinely not understand what you would get from an investment in your development. You need to explain the value to the company and reassure your manager that this activity won't impact your work commitments.

You need to find out your strengths which often coincide with your interests. Early in my career, the company did testing of aptitude and interests that was designed to help management understand where I should go within the company. My test results said I should be a scientist or a postal worker. I can see the science part but I wasn't sure about the postal part. I did collect stamps as a youngster but was never much into repetitious work, and am so mild mannered that going postal is unlikely.

The test said the last thing I should be is a manager. This test result was confirmed about five years into my career when I became the group leader of four very unusual engineers during the construction, commissioning, and startup of a huge plant in Texas. One was an experienced but eccentric Irish fellow. He accomplished more than anyone but did not impress anyone with his people skills. When he threatened to walk off the job, I went over to his apartment to talk him out of quitting and found absolutely nothing more than a phone. No bed, no table, and no chairs. The other three guys were fresh out of school with no plant experience. One was a hippie doing his own thing. The second was a Dutch guy who asked why he had to do anything I said, even the most trivial stuff. The third was a guy who wanted to be anywhere but there. I got through it all but decided I was not meant to be a manager. The test got three out of four right: I should be a scientist.

Fortunately, the company had a Fellow program, which meant that you did not have to go the management route to achieve higher grade levels and higher pay. I retired as a Senior Fellow, the same grade level as a director. Theoretically, a Distinguished Fellow could be at the same grade as a company president. Some companies have a technologist program analogous to a Fellow program. The technical and management routes are completely different. I think the technical route is more under your control and is surprisingly less stressful.

The clincher was when around 1990 our CEO said, in effect, "The company has no obligation to you and you need to be marketable just like this company's products." Some were shocked and dismayed. I felt enlightened. I appreciated knowing what was to come in terms of corporate culture.

Concept: Find out what you want to do in your career and take advantage of any training courses, conferences, resources, and tools that will increase your knowledge and allow you to distinguish yourself.

Details: Get last year's presentations and papers from conferences pertinent to your career and select the conferences you want to go to. For example, User Group conferences are great for getting in-depth information on the more effective use of the products from a supplier. The Automation Week conference is vendor neutral and has both technical and executive topics. The American Institute of Chemical Engineers (AIChE) conferences are intensive in process knowledge, and their relevance is excellent. If you have a degree in Chemical Engineering, you could build more process knowledge into your control systems. AIChE speakers such as Cecil Smith and Bill Luyben are exceptional in showing how to take advantage of process knowledge. Learn all of the tips in this book, especially the ones most relevant to personal development in your areas of aptitude and interest. Invest time in learning to use new tools such as asset management systems, instrument maintenance systems, model predictive control, tuners, and virtual plants.

Watch-outs: The Automatic Control Conference (ACC) is principally a forum for professors and their graduate students to present research. Unfortunately, university research and process industry practices appear to be from different universes, and, based on a read of the Proceedings, participation by industry technologists is declining. The Institute of Electrical and Electronic Engineering (IEEE) conferences are almost as academic and don't reflect much of an understanding of process dynamics, instrumentation, and process disturbances in industry. These conferences have academics talking to academics.

Exceptions: Don't expect to be able to go to more than one conference or training course a year. Expensive unproven software can be your undoing. Expert system software was largely a costly, dismal failure. The methodology was ad hoc and the interrelationships difficult to decipher. Expert system specialists were largely spinning their wheels and have little marketability today. Realize software is a tool and not an engineer. Recognize that to get the most out of tools requires analysis of what the tool is doing. Data analytics (multivariate statistical process control) may have an advantage over neural networks by offering an understanding of the relationship between inputs and outputs by principle components and contribution plots as noted in the 2010 February-March Control Talk column series "Drowning in Data, Starving for Information."

Insight: Making yourself marketable takes initiative, but it is beneficial to you and the company.

Rule of Thumb: Go to a technical conference or training course each year, and invest time in other ways that are in line with your aptitudes, interests, and goals. Enhance your marketability through increased knowledge and professional recognition.

Tip #59: Document the Benefits of Improvements

Greg McMillan (Career)

The process control improvement program described in the June 2012 Control Talk column "The Human Factor" was a success at the company I worked for because of one person, Glenn Mertz. Glenn was able to get agreement from plant management on the benefits achieved – no easy task.

Process efficiency and capacity vary with maintenance, season, days versus nights, operating points, raw materials, recycle streams, and operators. Mechanical and process engineers work to improve the equipment and the process. Operations tries to come up with better operating procedures. If there is an improvement, these people tend to get the credit because they have a better process or operational understanding and are closer to the action than are the automation engineers.

By way of illustration: In a batch reactor where an override control system was fixed and the key controllers were tuned, the process engineers took credit for the reduction in batch cycle time and the more consistent batches. The process engineers had more credibility than the automation engineers. After all, the details of override control, split range, back calculation connections, and the whole tuning thing are confusing at best to the people who are making decisions on the source and quantification of benefits.

How, then, can you as an automation engineer get well-deserved credit? First, you need to develop a good working relationship with process technology, operations, and maintenance (Tips #51 and #52). In particular, create a synergy between your skills and those of the process engineer supporting the unit. Get trend charts of before and after, ideally with metrics or at least with key process variables. In the case of the batch reactor example, a trend with feed rate, temperature, and batch time for the batches immediately before and after the configuration fixes and tuning would have provided the necessary evidence. Piping and equipment changes take much longer to implement than process control improvements. An operating point change resulting from a setpoint change can be fast but tests often need to be done and management-of-change procedures followed. In addition, the chance that an operating point change is made for the same batch that the override control system was fixed would be unlikely.

Unfortunately, once an improvement is made you can't undo it to show conclusively that the benefits were due to an improvement in process control. You can, however, use a virtual plant to show the "before" and "after." If the virtual plant can be made to match the process, the case will be more convincing when you turn off the process control improvements. You should run and adapt the virtual plant to be in sync with the real plant and speed up scenarios as necessary for your demonstration (Tip #99).

You have to be proactive, and prepare for people trying to take credit that should be yours. In the end, some negotiation may be required and the calculated benefits split up. If you are working with a process engineer, you should assign most of the benefits to him or her so you both look good, fostering future opportunities to work together. Success is a team effort and you would be wise to be generous in sharing the credit (Tip #57).

Concept: Getting proper credit for process control improvements requires snapshots and an analysis of process performance immediately before and after the improvements, along with a generous sharing of credit with the team, particularly the process engineer. A virtual plant can be used to demonstrate the improvements attributable to control improvements.

Details: Work closely with process engineers and operators to fully understand the problem and solution. Compute process metrics online, such as total raw material and total energy use for a batch or a shift (Tip #61). Get trend chart results of metrics and key process variables immediately before and after the process control improvements. Write a report briefly describing the improvements, benefits, and team effort. Review the report with the team members and your management and then ask for permission for it to be presented to the management of the team members.

Watch-Outs: Management of other disciplines may react negatively to the assignment of benefits outside of their section. Sharing the credit and getting the buy-in of team members can help mitigate this potential backlash. Remember also that shifts perform differently. Try to compare shifts with the same operators to help eliminate the variability.

Exceptions: Once you gain the confidence of plant management, you may be able to proceed immediately to the next improvement.

Insight: You need to have documented proof of the benefits of process control improvements in order to get the resources to advance your skills and expand your horizon to innovations instead of simply doing what has been done before.

Rule of thumb: Capture and report trends of key process variables and metrics immediately before and after process control improvements.

Tip #60: Write and Present Papers

Greg McMillan (Career)

In Monsanto Engineering Technology (ET), where I spent most of my career, I had the privilege of working with the world's best in modeling and control. The ET director, the late Dr. Jim Fair (Professor Emeritus at the University of Texas) encouraged us to document our research and development and to advance the technology of our profession by publishing the results. I didn't realize at the time how unusual this was, in that many engineers today are not encouraged or supported by their company to do this or are restricted because of concerns about the proprietary, legal, and public relations aspects of publication. Since writing is not a normal part of the job for engineers, the task can be formidable.

For me, the toughest part is getting started. Once I get past the first page, I get into a flow. I find an incredible release in expressing myself and being myself. Writing also gives me a sense of closure, allowing me to move on to new areas and opening up new perspectives.

I start with a Table of Contents with major subject headings. While a more detailed outline is probably a good idea for most people, too much detail up front suppresses creativity for me. I like the feeling of being free to let the flow take me where it needs to go. I find ideas evolve synergistically. Writing is an adventure for me with unexpected paths, sites, and discoveries. This book started with a title and 2-3 sentences of key points for each tip. We had a definite format in mind but no outline. I started each tip with a freewheeling introduction. When I sat down to write each tip, I often did not know what I was going to say.

It wouldn't work for everyone, but all of my publications have been written listening to classic rock with headphones. The music seems to inspire creativity by exercising the side of the brain that's opposite to analysis and logic. I find random play more stimulating, although most of my columns for a period of five years were written listening to Concrete Blonde. In the early years, I had to wait until I got into a feeling of the ideas flowing naturally. Now I use music as the inspiration from the start.

I write so ideas are not lost. I realized in the ET environment that a lesson learned is a lesson to be shared. To me, it is crime to see the knowledge that is being lost each day as the most experienced die or retire, never having gotten the opportunity to pass on knowledge gained the hard way. The automation profession is a school of hard knocks. During my college years, the lone course or two on control in Chemical, Electrical, and Mechanical engineering was steeped in math that 99% of automation engineers would never use on the job. We learn the most from making mistakes (ideally caught by mentors before the plant suffers). The sad fact is that without a source of knowledge, the same mistakes are made again and again, and nearly every new engineer goes through this rite of passage. My motto to this day is "Learn and Share."

In terms of career growth, what you publish serves to show what you have accomplished. This evidence was critical to my advancement to the Fellow level, my induction into the Process Automation Hall of Fame, and my attainment of the ISA Life Achievement Award. On any scale, your publications are the public record of your achievements. This record improves the marketability of your skills, which is increasingly important as companies change in profitability and direction: my original employer (3rd largest chemical company in the USA in the 1970s) became a life sciences company and ultimately a seed company.

Concept: Most of the knowledge required for successful automation applications in the process industry is undocumented. The implementers and maintainers of control systems are the best source of this knowledge, but they often don't perceive the benefit of publishing it as much as a supplier does in promoting a product. Publication of field experience promotes the growth of the individual and the profession. Share the recognition (Tip #57), be proactive in seeking approval of publications and presentations (Tip #58), and document the benefits (Tip #59).

Details: Try explaining what you know to a friend. Then write what you just said. Your friend can help. You might even make the friend a coauthor. You can try recording your conversations if this does not cramp your style. Break up the paragraphs and sentences (short is better). I prefer to present the concept and then the particulars. Most important, get started and don't over-think the process. Don't worry about the grammatical details at first. Writing is not an exact process like engineering. Just express your thoughts. Engineers tend to be overly concerned with exactly how to say something, resulting in a brain freeze when they are trying to write. You will learn from the edits by the copyeditor how to improve your writing. With practice comes skill. Remember, an idea imperfectly written is vastly better than an idea never written. For more on this, see my May 28, 2009 modeling and control blog "What Have I Learned – Writing."

Watch-outs: Avoid the first person ("I") except in prefaces and introductions. Avoid the use of "this is" and "it is" because readers may not be sure of what "this" or "it" is. Be careful to prevent the disclosure of proprietary information. Omit specific chemical compound names and operating conditions. Block out scale ranges of trend charts. When you present your paper, the worst thing you can do is to read from the paper. Slides should have brief bullet points, not paragraphs. Slides often don't have enough graphics. The only slides without a figure or trend chart should be introductory, procedure, overview, or summary slides. Even clip art is better than no art. A picture is worth a thousand words and is the key to keeping people from nodding off.

Insight: As an engineer, your publications are as important for the recognition of your expertise as the publications of a supplier are for the recognition of a product's capability, and your publications have an even greater potential to advance the automation profession.

Rule of Thumb: Write and present papers at ISA Automation Week using a supplier as needed to help you write the paper.

Tip #61: Install Online Process Metrics

Greg McMillan (Technical)

I never have quite understood why every raw material, utility, vent, reactant, recycle, and reagent flow rate and total is not ratioed to the product flow rate and total for each unit operation. The operator screens should display flow ratios (e.g., kg/kg or lb/lb), cost ratios (e.g., Euros/kg or $/lb), production rate ratios (e.g., kg/hr or lb/hr), and profit ratios (e.g., Euros/hr or $/hr). I have heard that major ethanol and pet food manufacturers do this, and have seen an increase in competitiveness between plants and shifts. I have seen energy ratios for boilers and kilns (e.g., kJ/kg or BTU/lb) but not much in terms of ratios for chemical and pharmaceutical plants. Knowing how process performance changes with changes in operators, maintenance, and process technical support may be disconcerting at first but ultimately productive, possibly spurring competition between operators, engineers, technicians, and plants to do better. Here the old adage applies, "You cannot control something you don't measure." In order to improve the performance of each unit operation, you need to measure the performance of each unit operation. As with anything measured, automatic control is better than manual control. A linear program (LP) in model predictive control (MPC) can do automatic optimization of the metrics by the use of cost ratios and profit to find the optimum intersection of operating conditions.

Nearly every process input that is set by operators or automatically manipulated by controllers is a flow. There should be a measurement of every flow for process analysis besides metrics. We are accustomed to flow measurements of raw material, reactant, and product streams, but wireless transmitters and insertion-type flowmeters (e.g., annubars) make flow measurement affordable for the remaining streams.

If we have all the flows, we can do a material balance and implement plantwide feedforward control (Tip #101). We can correct for pressure disturbances and valve nonlinearities by flow control, making the job of PID control (and particularly model predictive control) much easier. Because process gain is a nonlinear function of the ratio of manipulated flow to total flow, the process time constant is proportional to residence time, and transportation delay is inversely proportional to flow rate, we can intelligently schedule tuning settings. We can add wireless temperature measurements and do energy balance, heat release, and heat transfer calculations. We can develop inferential measurements of concentrations using neural networks, projections to latent structures, and first principles. We can improve the fidelity of a virtual plant. Since we have flow control, we do not need to have a pressure flow-solver in the virtual plant to know the flow through valves, and can adapt model parameters based on flow ratios (Tip #98 and 99).

Concept: Operating efficiency can be computed from a ratio of flows and assigning dollars per unit flow. Online metrics open the door to process understanding and innovations by the quantification of benefits and first principle relationships.

Details: For continuous processes, compute the increase in production flow or on-stream time or the reduction in the ratio of a utility, raw material, recycle, or reagent flow to a production flow. You should compute the ratios on a filtered instantaneous basis and as a ratio of totals for a representative period such as a shift. For batch processes, compute the ratios as flow totals to product total per batch and estimate the batch cycle time to get a production rate. For batch processes, efficiency is increased by higher batch end point concentration or lower total utility use and raw material feed per batch. Batch process capacity is increased by a shorter cycle time. Use ballpark estimates of dollars per unit mass (e.g., lb or kg) and normal production rates to get to the bottom line ($/hr or $/batch). Coriolis meters provide the ultimate in terms of accuracy of flow and density measurements and two-component composition measurements (Tip #73). Total heat release measurements can provide an inferential measurement of reaction conversion and the heat release rate can provide an indication of conversion rate and batch completion.

Watch-outs: Signal filters may be needed to reduce the noise in flow and cost ratios. To synchronize an upstream flow or utility flow with a downstream product flow, flows may have to be passed through a deadtime block and filter block to simulate transportation delays and residence time lags. There may be an inverse response and a temporary decrease in efficiency from the action of an LP and MPC that can cause impatient operations personnel to think the advanced process control (APC) system is doing the wrong thing. For example, if a reactant feed is increased to be closer to the optimum stoichiometric ratio, the yield would decrease if the change in reactant flow is not delayed and lagged to match the change in measured product concentration and flow out of the reactor.

Exceptions: The synchronization of raw material flows with final product flows after many batch and continuous processes may not be possible, causing metrics to be erratic. Synchronization is particularly difficult when there are several unit operations between a flow being manipulated and a product flow being measured for optimization.

Insight: Process metrics depend upon flow measurements.

Rule of thumb: Add flow measurements to every important stream and compute online metrics for the process efficiency and capacity of each key unit operation.

Tip #62: Demonstrate and Prototype Improvements via Dynamic Models

Greg McMillan (Technical)

During my career, modeling has been essential part of developing a better understanding of opportunities and a better control system. I have built simulations in IBM's Continuous Simulation Modeling Program (CSMP), Raytheon's Advanced Continuous Simulation Language (ACSL), Microsoft's Visual Basic, Hyprotech's HYSYS, Emerson's DeltaV, and MYNAH Technologies' MiMiC. In most cases I focused on the unit operations where control system innovation was possible and beneficial. I tackled all types of unit operations except for distillation. I avoided this unit operation because my associate Terry Tolliver was the world's best in simulating and controlling all types of columns[26]. My initial applications were in the modeling and control of difficult compressor surge control, furnace pH control, and strong acid-base pH control. In these applications the process response was so fast and/or extreme that simulations were the only way to safely analyze the response and test a solution[27,28].

One compressor I modeled could accelerate to the point of rotor damage within a few seconds once in surge. Because the compressor was responsible for all of an intermediate's production and rotors were one of a kind, the compressor could not be allowed to go into surge. A derivative module was installed that detected a high rate of change of speed and shut down the compressor before axial thrust was even measured. Simulation was the key to understanding how compressor surge was an incredibly fast phenomenon, like falling off a cliff. Compressors in two other applications were not as difficult to model but operations would often trigger a surge and have to shut down the compressors that were being manually brought online. In both cases, Operations insisted that automatic start-up was not possible because manual start-up was so difficult and unpredictable. I prototyped and demonstrated by dynamic simulation the automated start-up of the compressor flow, pressure, speed, and surge controllers. The go-ahead was given and the control systems thereafter provided faster and much more consistent start-ups with no surge[27,28].

For difficult control problems, I built dynamic first principle models that integrated ordinary differential equations (ODE) for charge, component, energy, material, and momentum balances as documented in Appendix F. Today, for common unit operations you can graphically drag and drop in modeling objects instead of writing and programming the ODE.

Via a test, an auto or adaptive tuner, model predictive controller, or rapid modeler module can identify the dynamics for a given combination of a process input and output. The test typically consists of a momentary step change in a flow, often by a step change in a PID setpoint or output. The identified deadtime, process gain, and process time constant can be used to provide an experimental model. This model is linear and may not reflect all the interactions and effects of operating conditions, but it may be more accurate than the first principle model for operation at the test conditions.

Concept: Experimentation on actual processes is limited at best. Changes in operating points often require documentation and approval. New control systems must improve and not hinder plant performance from the "get-go." First principle and experimental models can provide a virtual plant to develop and demonstrate new strategies and provide implementation details and expected benefits.

Details: Building and using first principle models based on the conservation of energy, mass, and momentum will help you to develop a deeper understanding of process interactions and dynamics. Process Flow Diagram (PFD) types of dynamic simulators excel at capturing the complexities of process relationships and are useful for detailing the effect of process conditions on process gains[26]. Advanced Modeling Objects (AMO) are more easily integrated than PFD simulations into a virtual plant and offer more realistic dynamics by the inclusion process deadtime, batch and startup responses, and automation system responses. Experimental models obtained from auto and adaptive tuners, model predictive controllers, and rapid modeler modules offer accurate process dynamics for the test conditions. See the May 6 and May 11, 2012 *ISA Interchange* posts "**What are the types of Dynamic Simulators and their pros and cons?**" and "**What is the best approach in developing a dynamic process model?**" for more details on the relative merits of simulators, fidelity levels, and the steps for constructing first principle models for the Virtual Plant described in Tip #99 and the "Checklist for a Virtual Plant" in Appendix C.

Watch-outs: Steady-state models are only useful for finding gains of self-regulating processes. Most PFD simulators do not extend equilibrium relationships to the driving force equations needed for dynamics. PFD dynamic simulators are notorious for having insufficient deadtime in the model due to the lack of transportation delays, mixing delays, valve backlash and stiction, sensor lags, measurement update times, and signal filters. Nearly all first principle simulators treat volumes as completely mixed and ignore injection delays from dip tubes. For volumes with less than perfect mixing and small reagent or reactant flows, the process deadtime is an order of magnitude or more too low. For fast processes such as liquid pressure and flow, modeling transmitter damping, scan rates, and PID execution times is crucial for getting the dynamics right. Experimental models can break down due to process interactions and nonlinearities. Experimental models are generally not valid for start-up, shutdown, and abnormal conditions. Backlash and stiction must be modeled to show limit cycles and loop deadtime as the PID output signal works through the deadband, resolution, and threshold sensitivity of the control valve. The PID algorithm in a distributed control system (DCS) is sophisticated and proprietary, representing many engineering years of effort. The PID emulated in dynamic simulators generally does not capture all of the features of the DCS PID. In particular, external-reset feedback, a powerful tool, must be included[4]. For these and many other reasons, the use of the actual DCS configuration in a virtual plant environment (Tip #99) is essential.

Exceptions: Simulations involving computational fluid dynamics and partial differential equations are presently beyond the capability of the automation engineer.

Insight: First principle models and experimental models used in a virtual plant offer a powerful tool for process control improvement that can be extended into an operator training system.

Rule of Thumb: Develop a dynamic model for unit operations where process control improvements can translate into significant plant performance benefits.

Tip #63: Use Field Analyzers to Measure Key Component Concentrations

Greg McMillan (Technical)

When we did process control improvements in the 1980s and 1990s, the major limitation was the lack of a reliable field analyzer. None of the plants had field analyzers on raw materials. The specialty chemicals production units had very few field analyzers and were flying blind. Plants for chemical intermediate products had field analyzers on most key product streams that were supported by an excellent, extensive plant analyzer group. Unfortunately, the technology was old, dating back to the 1970s. Many of the analyzers, analog circuits, and sampling systems were literally homemade and were dependent upon the expertise of the analyzer group. Most of these analyzer specialists have retired. Fortunately, the technology developed was adopted by new analyzer companies. Subsequently, the technology has steadily advanced and the electronics have become small computers providing diagnostics, intelligent interfaces, and standardized communication. Sample valves and sampling system design have also improved. Analyzer systems are more reliable today but still require maintenance and special expertise. Unfortunately, onsite analyzer specialists are becoming extinct.

Quite a bit of effort was devoted in the 1990s to developing artificial neural networks (ANN) to predict compositions in streams instead of installing analyzers. Unfortunately, the technology was oversold by big neural network suppliers, who claimed, "No hardware, no engineering, and no maintenance. Just dump all of your plant data into the ANN." PID setpoints, process variables, and outputs were inputs, ignoring the fact that the PID algorithm was in play. The necessary Design of Experiments (DOE) was not done. Steady-state data without changes to process inputs led to relationships that violated first principles and wild extrapolations when the plant deviated from the test conditions. Principal components and drill-down into contributions were not available. Process engineers did not review the relationship of each input to the predicted output. Automated feedback correction from lab analysis was often not used. As the result of all this, ANN's achievements were temporary at best. Today, ANN integrated into a DCS are easier to use and may offer benefits for focused applications.

There are some examples of a brighter future for online and at-line analysis. Coriolis meters offer an incredibly accurate, reliable, and nearly maintenance-free density measurement for online analysis where two components have significantly different densities. The capability has been extended with more sophisticated digital computations to include percent solids and bubbles. Conductivity and pH offer online analysis for high and low concentrations, respectively, of acids and bases. The Nova BioProfile Flex at-line analyzer with an automated sampling system is becoming the standard for bioreactors in the biopharmaceutical industry. Within minutes, the Nova analyzer can provide cell size, count, and health along with the concentration of nutrients and inhibiting byproducts from 1 ml samples.

Concept: Process efficiency and capacity often depend upon the composition of input and output process streams. Automation systems commonly measure temperature, pressure, level, pH, and flow but rarely composition. Sometimes pH, pressure, and temperature can be an indicator of composition in reactions and separations for a given set of input stream compositions. At present, lab measurements are relied upon for confirmation of the estimated values. Large continuous plants tend to have analyzers on product streams. What is primarily missing despite advances in technology are on-line and at-line analyzers for raw materials and batch operations.

Details: A history of lab results can show the variability of key components in key process streams. Simulations can show the effects of changes in composition on the process. Field analyzers on streams where the key components show variability or need to be optimized can be used for a higher level of control. The higher level can be as simple as cascade control or as sophisticated as model predictive control. Coriolis meters should be installed on all reactant and product streams (Tip #73). Online analyzers require less maintenance than at-line analyzers because there is no sampling system but they often provide an inferential measurement that does not reflect the effects of changes in process conditions. The most effective analyzer for the process industry is the gas chromatograph, as discussed in the December 2011 Control Talk column "Analyze This!" and the January 2012 Control Talk column "Gas Chromatographs Rule."

Watch-outs: Droplets or solids remaining in process samples after sample conditioning will adversely affect the results from a gas chromatograph (GC). The sample used in a GC must be completely vaporized. The sample point for any at-line analyzer may not be representative of the process composition because of the separation of phases and non-ideal mixing at the point of sample extraction. Near Infrared (NIR) analyzers are only as good as the set of samples used to develop the Projection to Latent Structure (PLS) statistical models. NIR models (NIR calibration) must be updated as raw materials and operating conditions change. Special mathematical expertise is needed for understanding and improving NIR models. The maintenance costs of analyzers (except for Coriolis meters) usually exceed the hardware cost. Analyzers should not be used in closed loop control until they have proven to be sufficiently accurate and reliable in actual plant operation. At-line analyzer sample time, cycle time, and multiplex time will increase the total loop deadtime, destabilizing the loop, unless the controller is retuned or an enhanced PID developed for wireless is used. Online first principle models and experimental models (e.g., linear dynamic estimators) periodically corrected by at-line analyzers can provide an immediate predicted composition, eliminating the additional deadtime[29,30].

Exceptions: Analyzers should not be installed if there is no onsite support with the required expertise or if the Return on Investment (ROI) is insufficient based on actual analyzer downtime and life cycle cost, which includes maintenance cost.

Insight: Field analyzers enable a higher level of control, such as model predictive control, to improve product quality and process efficiency and capacity.

Rule of Thumb: Install a field analyzer on key continuous streams and batch unit operations if there is an adequate ROI and onsite technical support.

Tip #64: Improve Setpoints

Greg McMillan (Technical)

The classic case for process control improvement uses a figure that shows the mean and standard deviation of a statistical distribution and the optimum of a process variable (PV). The optimum is often taken as the constraint on product quality. The case is made that if PV variability is reduced, the setpoint for the PV can be moved closer to the constraint. In my experience the bias to the setpoint due to variability is less than the bias by the operator because of measurement error, poor measurement and valve turndown, manual actions, on-off actions, abnormal operation, unknown changes in raw materials, equipment deterioration and modifications, process modifications, lack of process knowledge, operator "sweet spots" and war stories, and tactical tradeoffs between capacity and efficiency.

In this tip we will address the automation system considerations of improving setpoints, but there are bigger questions such as integration of the knowledge of the effects of equipment, piping, and process conditions and changes into the decision to move the setpoint. Simulations can provide the process knowledge. Unfortunately, maintenance and operational databases have not achieved the level of integration needed to understand how a setpoint should be changed, as noted in the April 2010 Control Talk column "Drowning in Data, Starving for Information - Part 4."

You may just need to move the setpoint. Better process knowledge or a simple trial run at a lower or higher setpoint might be sufficient. I have seen cases where running models provided the knowledge and confidence to try a better setpoint. In other cases, big improvements were directly made by engineers with practical process and control capability. Simply entering a better setpoint may get you a bonus, or at least a free meal.

A common example is temperature control. Distillation columns rely on temperature for an inferential measurement of composition. Temperature determines both production rate and quality in reactors. Reaction rate increases with temperature but just past the optimum, reverse reactions, side reactions, and product degradation may occur. Consequently, there is a peak in the plot of production rate versus temperature. For bioreactors, the specific cell growth rate and product formation rate increase with temperature but cells start to shut down and die at temperatures to the right of the peak in the plot of specific growth rate versus temperature. The peak is different for cell growth and product formation, so temperature shifts are made generally after the exponential cell growth phase is fully established. However, the best size and timing of the shift are often not known. A similar situation exists in bioreactors for pH, with an even sharper peak. Measurement accuracy and control requirements of a few hundredths of a pH of setpoint are stated. The question is, do we really know the best setpoint to the same degree of precision?

Concept: Unless off-spec product is being downgraded, rejected, or recycled, the benefits of process control improvement are not realized until a setpoint is moved closer to the optimum. Greater plant knowledge offered by simulation and the integration of databases can lead to more intelligent setpoints. Better control strategies, measurements, control valves, and tuning and higher levels of control can enable operation closer to a constraint.

Details: Eliminate instrumentation and valves as the cause of a bias between the setpoint and the optimum. Coriolis meters, magnetic flowmeters, and precision throttling valves with sufficient pressure drop can eliminate limit cycles and errors in the ratios of flows and improve plant turndown. Resistance temperature detectors (RTD), integral mounted temperature transmitters, direct mounted pressure transmitters, and radar level indicators reduce measurement errors. The latest improvements in sensor, transmitter, and positioner technology can eliminate setpoints being shifted due to automation system limitations. Develop plant knowledge to find more optimum operating points. Process simulations, integrated maintenance and operation databases, and an asset management system (AMS) facilitate finding better setpoints. Use a higher level of control to automatically find better optimums. Analyzers (Tip #63) offer a higher level of control. Feedback loops that fully exploit smart PID features (Tips #91-96 and #100) can eliminate operator actions (Tip #69), on-off actions, abnormal operation, and activation of safety instrumented systems (SIS). Valve position control (VPC) strategies described in Tip #97 and model predictive control (MPC) and a linear program (LP) can automatically optimize the setpoints of unit operations. Real-time optimization (RTO) can provide the ultimate in optimization of setpoints throughout a continuous plant if there is an accurate model and sufficient RTO expertise onsite.

Watch-outs: Operations personnel may be reluctant to believe that manual actions can be eliminated or that operator sweet spots and war stories are not valid reasons for operation further from a constraint. If there are no online process metrics (Tip #61), operators will naturally choose the setpoint that minimizes any perceived potential disruption, stress, and extra work. If the goal is business as usual rather than a more profitable business, Operations will be reluctant to make changes. If the operators are not fully trained in the higher levels of control (Tip #99) or control room support is not provided for all shifts for a sufficient duration, new control systems will be put in manual whenever something unusual happens.

Exceptions: If there are large blend tanks to attenuate product variability and if process capacity rather than process efficiency is the goal, flows may simply be set at a maximum. For conventional rather than fed-batch operations with only on-off valves (no control valves), the optimization involves batch sequences and totalized batch charge flows rather than setpoints.

Insight: Setpoints can be improved by achieving a higher level of control and knowledge.

Rule of Thumb: Use automation systems and plant knowledge to operate closer to constraints.

Tip #65: Use High Rangeability Flowmeters and Valves to Increase Plant Turndown

Greg McMillan (Technical)

I became sensitized early in my career to flowmeter and valve rangeability when I inherited the responsibility of solving pH control problems from the late Vernon Trevathan. The 0-14 pH measurement scale covers 14 orders of magnitude of hydrogen ion concentration (14 orders of magnitude of strong acid or strong base concentration). Standard practice in the 1970s dictated the use of one neutralization stage for every 2 pH units away from the control band (allowable error around the pH setpoint). For a 7 pH setpoint, neutralization of influent pH at the extremes of the scale required three stages of neutralization with successively smaller valves, each with a real rangeability requirement of 100:1. These exceptional requirements led me to find that the real limitations to rangeability are important for nearly all applications.

I quickly learned that available details on flowmeter and valve rangeability were missing or misleading. Some suppliers of on-off rotary valves posing as control valves claimed a rangeability of 200:1 based on the ratio of the maximum to minimum flow coefficient (C_V) of the inherent flow characteristic. Another definition based on deviation from the inherent characteristic led to the conclusion that linear trim was best. The real rangeability for a valve is the ratio of the maximum controllable flow to the minimum controllable flow, taking into account valve precision and the slope of the installed characteristic.

Real rangeability depends upon valve backlash and stiction near the closed position and the percentage of the system pressure drop allocated as control valve pressure drop. The minimum Cv is more realistically the Cv for the deadband, resolution, and threshold sensitivity. For example, if the resolution is 0.5%, the minimum Cv is the inherent characteristic Cv at 0.5% signal. Stiction and consequently resolution and threshold sensitivity are often worse near the closed position, the result of seating and sealing surfaces. Supplier-provided estimates of deadband, resolution, and threshold sensitivity are for 50% signals. The minimum controllable flow is then computed from the minimum Cv and the ratio of valve drop to system drop. Equations 7-19a through 7-19d in the ISA book *Essentials of Modern Measurements and Final Elements*[11] provide the simple series of computations needed to provide a practical estimate of valve rangeability.

Sliding stem control valves designed for throttling service with diaphragm actuators and digital positioners generally have the least backlash and stiction. Fortunately for pH control, the reagent flows are so small that a sliding stem valve is economical and the valve pressure drop is more than 50% of the system drop due to low frictional losses in the piping system from low flow.

I also discovered that the real rangeability of differential head meters and vortex meters depends upon the signal-to-noise ratio, which in turn depends upon the piping geometry. Fluctuations in the velocity profile, and particularly swirling, from upstream valves, fittings, elbows, and changes in plane increase the noise. I found that manifold taps and flow nozzles or venturi tubes and straightening vanes offer less noise than orifices with single taps and just straight runs. The addition of a low range differential pressure transmitter can extend differential head meter rangeability if measurement noise is minimized.

Concept: Plant turndown depends on the ability to control the process at low flow rates. Valve and flowmeter rangeability statements in the literature are often optimistic and don't take into account practical limitations such as backlash and stiction for valves and signal-to-noise ratio for flowmeters.

Details: Valves with minimum packing, seating, and sealing friction have the best threshold sensitivity and resolution (precision). Sliding stem valves with a direct connection of actuator shaft to plug stem have the least backlash and least friction. Diaphragm actuators and digital positioners have the best threshold sensitivity. The trend to allocate a smaller percentage of the system drop to valve drop to reduce energy use does not consider the accompanying reduction in valve rangeability. The drive to reduce upfront hardware costs does not take into account the loss in real rangeability. Flowmeters with the least sensitivity to fluid velocity profile have better rangeability from a better signal-to-noise ratio at low flows. For flowmeters with good signal-to-noise ratios, the threshold sensitivity becomes the limiting factor to rangeability. On the basis of these considerations, properly sized Coriolis meters and magmeters have a rangeability that is 20 and 10 times, respectively, better than the rangeability of the best differential head and vortex meter installations.

Watch-outs: If a valve manufacturer does not know what is meant by the terms deadband, resolution, and threshold sensitivity, the valve was probably originally designed for on-off applications and is not suitable for throttling service. Statements that only 5% of the system pressure drop needs to be allocated to valve drop do not take into account the loss of turndown and the distortion of a linear inherent characteristic into a quick opening installed characteristic. Very low flow rates and viscous flows can cause the Reynolds number to go from the turbulent to the transition region. In these applications, a roller diaphragm valve that forces laminar flow should be considered. See tips #77 – #85 for more details on poor valve package and system design. Vortex meter signals will become erratic if the velocity is too low. The signal must be forced to zero. An oversized vortex meter, where the maximum meter flow is much greater than the maximum process flow, can cause an unexpected drop out of the flow signal.

Exceptions: If a continuous process runs at one steady-state production rate, and there are no automated start-ups or transitions, turndown may not be a concern.

Insight: Flowmeter signal-to-noise ratio and valve backlash, stiction, and pressure drop determine the turndown capability of a plant.

Rule of Thumb: To increase plant turndown, use control valves and flow measurements with the best low flow response and precision.

Tip #66: Seek First Principles for a Deeper Understanding

Greg McMillan (Technical)

The automation profession requires knowledge of thousands of details on hardware, software, interfaces, and process applications. The diversity of the industries served and the natural inclination of people to think their case is special often lead to the conclusion that each plant and application is unique. Automation engineers are particularly vulnerable because of overwhelming details to "not being able to see the forest because of the trees." Maybe because of my orientation as a physicist, I am always trying to find some commonality via a concept based on a first principle. I use applications of the principle and user feedback to either confirm or update the concept, an essential part of the scientific method and the reason that humans have advanced compared to animals (except for the special case of fraternity house beer blasts).

A good example of the power of concepts involves deadtime (Tip #70). With the guidance of Greg Shinskey, I found that PID tuning and errors for load disturbances could be simplified to simple functions of deadtime (Tip #71). Hundreds of equations developed over the last 60 years reduce to these relationships when the objective is load disturbance rejection[1,2]. Then I realized that a near-integrating process gain could be universally applied if a deadtime function block was used to identify the process variable (PV) ramp rate[1,2]. The initial PV responses of self-regulating, integrating, and runaway processes all look like a ramp to the PID. The deadtime block even enabled the unexpected extension of the near-integrator ramp concept to loops where the loop deadtime is much larger than the time constant (as noted in the July 19, 2012 *Control Talk* blog "Deadtime Dominance Does Not Need to Be Deadly.")

Subsequently, I found that the same deadtime block could be used to compute the slope of a batch profile for the optimization of feed rate and cycle time[3]. I also discovered that the block could compute a future PV value for a full throttle batch and start-up response[4]. Greg Shinskey clued me in to the concept of simply using a setpoint filter with the time constant equal to the reset time to eliminate overshoot on setpoint changes, enabling the use of load disturbance tuning. My realization (that of external-reset feedback) enabled the use of direction setpoint rate limits to achieve a multitude of other process objectives, including abnormal situation prevention, coordination, and optimization with retuning (Tip #72). Furthermore, this external-reset feedback as part of an enhanced PID could, without retuning, handle large wireless and analyzer delays, stop limit cycles, and eliminate interactions. Control engineers can move on from tuning to taking advantage of a huge step increase in PID capability (Tips #89-97).

I did a batch project just to see how different the batch world is from the continuous world, which is the subject of university classes and most control theory textbooks. The essential aspects are an integrating process response from a closed liquid discharge valve during the batch and the use of sequences. I then realized that a continuous process can be treated as a near-integrator when running and treated as a batch during start-up.

Sequences used to automate batches could be used to automate start-ups of continuous processes.

Concept: Conceptual knowledge should be sought to see the commonality in seemingly different application requirements. Personal pride in the uniqueness of solutions should take a back seat to the elegance of simplification and unification, which will extend the capabilities of the automation profession. Conceptual knowledge should be supplemented with procedural knowledge as a guide for effective implementation of solutions based on the knowledge.

Details: Conceptual knowledge is best developed in the time domain from trend charts, thought experiments, and from the principles of material, energy, and charge balances in Appendix F. The August 3, 2012 *Control Talk* blog "Where do Process Dynamics come from" discusses how these balances can lead to a better understanding of process dynamics. For example, all of the individual ionic equilibrium calculations for pH in a half dozen textbooks can be simplified to a single charge balance equation that is solved by the simplest of all search techniques: "interval halving." Use the checklists in Appendix C to help provide procedural knowledge for effective implementation.

Watch-outs: You have to keep an open mind and avoid preconceptions. You should look for cases that provide exceptions to the concept along with cases that provide evidence that support the concept. The developers of new tools are (unintentionally or intentionally) often looking for proof that their creation is valuable. The use of tests to prove the value of an innovation is commonly seen in papers on control algorithms. Better tuning, tests on unmeasured disturbances as process inputs, and the use of key PID features may have considerably reduced the perceived advantage of an algorithm. Because loop performance depends on tuning (Tip #71), you can tune a controller to prove any point.

Exceptions: The first requirement is to get the job done right and on time. Flashes of insight that reveal concepts for unification of diverse and seemingly contradictory relationships cannot be predicted or scheduled.

Insight: Principle-based conceptual thinking frees you to move to a higher level of accomplishment.

Rule of Thumb: Seek to find the commonality rather than the uniqueness of solutions.

> **"Personal pride in the uniqueness of solutions should take a back seat..."**

Tip #67: Find the Best Measurement Location

Greg McMillan (Technical)

I became sensitized to the importance of measurement location when I found the easiest way to keep a pH electrode from fouling was to install it in a pipe with a flow velocity of 5 to 7 fps[25], preventing the usual 100× deterioration in the speed of response that resulted from just a few millimeters of coating. The higher velocity also made the electrode much faster-responding when clean. The conventional wisdom of putting an electrode into a vessel was proven wrong on several counts. The velocity in even the most highly agitated vessels is only 1 fps, resulting in a slow response and the need to remove the electrode more frequently. Furthermore, removing an electrode from a vessel in service is more problematic than removing it from a recirculation line that can be isolated.

I also found that pH electrodes installed too close to the outlet of a static mixer were too noisy. Moving the electrodes downstream 25 pipe diameters made a world of difference. The increase in loop deadtime was only 1.5 sec from the additional transportation delay (9 feet of 4-inch pipe at 6 fps). The decrease in noise allowed me to use a smaller filter time so that the actual total loop deadtime was less.

The same principle applies to thermowells, although the effect is less dramatic. Higher velocities decrease the fouling rate and decrease the measurement lag, the result of an increase in the heat transfer coefficient. (The annular clearance (air gap) between the sensor and the inside diameter of the thermowell has a bigger effect than velocity[24].) The thermowell should also be about 25 pipe diameters downstream of a heat exchanger to allow for mixing of the flows from the tubes.

Bubbles in liquid streams and droplets in gas streams cause noise when they hit a sensor. Bubbles from air, oxygen, and carbon dioxide spargers in bioreactors and chemical reactors can cause dissolved oxygen and pH signals to become noisy. Droplets of water at a desuperheater outlet cause a noisy temperature measurement. Ammonia bubbles at a static mixer outlet cause a noisy pH measurement.

The tip of an electrode or thermowell should be near the pipe centerline because the temperature and composition vary over the cross section of the pipe. For highly viscous fluids, the error is pronounced. I found that the temperature measurement in extruder outlets and the pH measurement in static mixer outlets with a sulfuric acid reagent are particularly sensitive to the depth of insertion of the sensor tip due to the effects of the high viscosity of polymers and of 98% sulfuric acid.

Differential head meters and vortex meters should be located where the velocity profile is uniform, the flow is turbulent, and there is a single phase - or wherever the piping designer tells you (just kidding).

Concept: The sensor location should provide sufficient residence time and mixing to ensure a single phase and a uniform mixture. The location should minimize the volume between the point of injection and the sensor to minimize delay. For differential head and vortex meters, a consistent velocity profile is required. Most importantly, the location must be sensitive to changes in both directions of the process.

Details: Maximize the detection of changes in the process from disturbances and setpoint changes. For composition, pH, and temperature choose the location that shows the largest change in both directions for a positive and negative change in the ratio of the manipulated flow to the feed flow, realizing that there are cross-sectional and longitudinal temperature and concentration profiles in pipes and equipment. For distillation columns, the best location for the thermowell is the tray with the largest change in temperature for an increase or decrease in the reflux to distillate ratio or steam to distillate ratio. A temperature or pH sensor and an analyzer sample tip should be near the center of a pipe and should extend well past equipment walls. A series of temperature sensors across a fluidized bed at several longitudinal distances is often necessary, with averaging and signal selection to get a representative measurement and prevent hot spots. The insertion length of a thermowell should be more than five times the diameter of the thermowell to minimize thermal conduction-induced errors from heat conduction along the thermowell wall between the tip and the process connection. To prevent vibration failure from wake frequencies in pipes, calculations should be run with the program supplied by the manufacturer on the allowable maximum length. A location with good mixing and a single phase will minimize fluctuations in measured temperature and concentration and the disruption caused by bubbles or solids in liquids and liquid droplets in gases hitting temperature or pH sensors or getting into sample lines for analyzers or into impulse lines for pressure and level measurements. Pressure probes in high-velocity gas streams and furnaces must be designed to minimize momentum and vacuum effects. Sensors and sample probe tips should not be installed on pump suctions and should be downstream of strainers. Minimize sensor deadtime and lag by reducing transportation delays and increasing velocities. The transportation delay in a pipe or sample line is the volume divided by the flow rate or the distance divided by the velocity. The lag time of temperature and pH sensors decreases with velocity by an increase in the heat transfer and mass transfer coefficient, respectively. Fouling also decreases with velocity.

Watch-outs: Material volumes behind baffles or near the surface or bottom of an agitated vessel or at the outlet of inline equipment may not be well-mixed. Packed and fluidized bed equipment may have uneven composition and temperature distribution from flow channeling. Programs for vibration analysis may only be looking at thermowell failure and will not predict RTD failure. The use of calcium hydroxide (lime) or magnesium hydroxide as a reagent may seem cost-effective until you consider the cost of poor control and solids going downstream.

Exceptions: The best location may not be accessible or maintainable due to height or obstructions.

Insight: The best measurement location maximizes the sensor sensitivity and maximizes the signal-to-noise ratio and minimizes deadtime.

Rule of Thumb: Find a location that is sensitive to changes in the process, where the fluid has a uniform mixture and a single phase, and where sensor lag and transportation delay are minimized.

Tip #68: Find the Best Valve Location

Greg McMillan (Technical)

When I am in an instrument and valve repair shop, I see many more control valves than instruments, particularly with the advances in sensor technology, transmitter intelligence, and asset management systems. Valves are mechanical devices and as such require more maintenance. Packings, seals, seats, and O-rings wear out. To ease maintenance, a control valve must be located so it can be readily and safely removed from the pipeline. However, there is much more to consider in locating a control valve.

With liquid streams, a change in control valve position causes an immediate change in liquid pressure and, in a full and pressurized pipeline, a pressure wave traveling at the speed of sound. In less than a second, the pressure imbalance from the wave provides a driving force that overcomes the liquid's inertia and accelerates the liquid to a new velocity. Consequently, within a plant the location of a control valve does not appreciably affect the response of a process variable (PV) within the pipeline. However, if you are measuring PV response in a pipeline and the control valve is in a different pipeline, throttling a flow that is being added to the pipeline that is being measured, the distance of the valve from the measurement may cause a transportation delay. This piping and valve arrangement commonly occurs in the dilution and blending of streams.

There are some really bad control valve locations that show no understanding of the negative impact of deadtime by process and mechanical design engineers (see Tip #70). One of the worst is where there are several vessels between the control valve and the measurement. The residence time of the smaller vessels in series with the larger vessel becomes deadtime. Also bad is gravity flow. Now a change in control valve position starts a wave traveling slowly down the partially filled pipeline. For small flows in vertical runs, the flow is a falling film causing a large and unpredictable deadtime. The worst case of deadtime resulting from valve location is encountered in pH control. A control valve often throttles the reagent flow to a dip tube. The dip tube has a minimum size for structural integrity and normal mixing rules put the dip tube down near the impeller. Unfortunately, this creates a dip tube volume of 2 gallons. If the reagent flow is 1 gallon per hour, when the control valve opens, it takes 2 hours for the reagent to flush the process fluid out of the dip tube. When the control valve closes, the reagent will continue to migrate into the process for several hours. The solution is to have the control valve add the reagent to a high flow recirculation line, thereby reducing the injection delay to seconds.

To prevent flashing, control valves should be located so the fluid pressure in the vena contracta (that is, the narrowest opening in the flow path) does not drop below the vapor pressure of the fluid. If flashing cannot be avoided, the valve type and trim design should be selected to prevent cavitation in the valve or downstream equipment. Stan Weiner, my coauthor of the Control Talk column, recommended the flashing control valve be installed directly on an inlet nozzle near the top of a vessel so the collapsing of bubbles would occur in the vessel vapor space when cavitation could not be prevented.

Concept: Control valves do not require much in the way of straight runs and do not introduce appreciable delays within plants when they are in the same pipeline as the measurement. (Long-distance oil and gas pipelines are another story.) Valves adding flows to process equipment can cause large injection delays or bypass mixing in the equipment. Valve location on streams to equipment should not introduce excessive deadtime or noise. Valve location, type, and trim should minimize flashing and cavitation and provide safe and easy access for removal and repair. On-off valve locations should minimize the volume to the destination when flow is stopped to prevent totalization errors in charges.

Details: Control valves should be at floor level or accessible from platforms. Block, flush, and drain valves should be installed to enable them to be safely removed. Control valves should be located on the same equipment or pipeline as the measurement and downstream of flow measurements. Reagent control valves should be moved from dip tube to recirculation line injection to eliminate injection delays for pH control. On-off valves close to the point of injection should be added to provide isolation and to shut off the flow. For pH and reactor control, the volume between the on-off valve and the nozzle should be minimized by flanging the on-off valve to the nozzle or nozzle block valve for low reagent and reactant flows and high process sensitivity. Gravity flow piping is not recommended because of variable head and velocities, but if it is used, the control valve should be as close to the nozzle of the destination as possible.

Watch-outs: The location of the nozzle and dip tube entry points into a vessel must not result in the flow being injected close to an exit nozzle; thereby short circuiting inlet flow to outlet flow and bypassing the mixing in the vessel. Throttle valves should not be used as isolation valves, and isolation valves should not be used as throttle valves (Tip #83). If an on-off valve for batch control is not close to the flowmeter, the pipeline inventory between the on-off valve and flowmeter can cause the charge to be significantly different than the batch setpoint. To minimize excess charge, the on-off valve should stroke as fast as necessary when the "close" command is given.

Exceptions: For Coriolis meters in liquid service with no possibility of flashing, the control valve can be located upstream of the flowmeter because a Coriolis meter is not sensitive to velocity profile.

Insight: Control valves can cause damage to piping from cavitation and poor control from injection delay and short circuiting.

Rule of thumb: Locate control valves to be maintainable, provide fast injection into mixing zones, and prevent flashing and cavitation.

Tip #69: Add Control Loops to Eliminate Manual Actions and Sequences[6]

Greg McMillan (Technical)

I once taught a course in process control to junior and senior chemical engineering students as an adjunct professor. Students were generally in the class not because they thought process control was important, but because the course was required to get a degree. One student asked me why you needed a control loop when all you have to do is set the flow to match what is on the process flow diagram (PFD). These students had almost exclusively been trained to understand and model steady-state processes.

When I was reviewing the opportunities for process control with a process engineer at a major pharmaceutical company, the engineer said he wanted to set the input flows. He would schedule the air, oxygen, reagent, glucose, glutamine, and nutrient flows to the bioreactor. He thought control loops were not desirable because they would vary these flows. While most other pharmaceutical companies use dissolved oxygen control to manipulate air and oxygen flow and pH control to manipulate reagent flow, the other flows are still scheduled even though analyzers such as the NovaFlex BioProfile (Tip #63) are now available for closed loop control. Going from traditional batch (where flows are charged sequentially) to fed-batch (where flows are manipulated by control loops) is a big step for most process engineers. Even tougher for process engineers (and operators) to understand is that control loops can provide automated start-ups and abnormal situation protection by the use of key PID features.

The basic principle missing in the education of process engineers and operators is the benefit of the transfer of variability from key process variables (PV) to input flows in the process industry. A control loop can do this through feedback and feedforward control to compensate for unknowns, varying demands, and unmeasured disturbances. Through intelligent guesswork and models process engineers try to set flows to achieve a desired PV. These engineers don't realize that plant models have thousands of parameters with errors. Processes are routinely subjected to disturbances from changes in raw materials and in the case of bioreactors, the whims of living cells. Batches inherently have changing demands. For example, the oxygen demand for a bioreactor can increase by two orders of magnitude from start to finish.

Humans are also bad at anticipating the effect of lags and delays on process response. The natural consequence of this lack of patience is to overreact. A PID controller will not. It has knowledge of the process dynamics, which is built into the tuning settings. In addition, even if humans had better knowledge and patience, they cannot possibly be as attentive as a control system. A PID constantly looks at results and makes corrections, consistently and without fatigue or distraction.

If a process engineer or operator says that manual actions or the scheduled actions are too complex for control, the application is an excellent candidate for complete automation via a PID controller. A control loop can replicate the best of these actions. The consistency of the automated response can lead to better recognition of how to improve these actions. Feedback control automatically takes care of unforeseen effects.

Concept: The best actions of process engineers and operators can be automated as the starting point for a control loop. The loop will subsequently compensate for unknowns, varying demands, unmeasured disturbances, and changes in desired operating points, opening the door to the improvement of actions and the optimization of setpoints.

Details: Work with Operations and process engineering to make sure you understand the process relationships and abnormal conditions. Set the initial manual actions for the start of a batch or unit operation and the best reactions for abnormal conditions via the output tracking mode of the PID. Hold the output long enough for the PID to see the response to the actions. Release the PID for feedback control (Tip #91). Add ratio control to enforce and optimize ratios (Tip #93). Add feedforward control to give fast, smooth transitions to different product grades and production rates for flexible manufacturing (Tip #101). Tune the PID and use key features (Tip #71) for good disturbance rejection and setpoint response. Use additional features to provide faster or slower responses based on direction and on coordination and optimization objectives (Tip #72). See the November and December 2009 Control Talk columns "Show Me the Money" for case histories on the elimination of operator actions. Use a virtual plant (Tip #99) for developing, prototyping, testing, demonstrating, and training. Keep the most experienced operator and process engineer involved from start to finish. Support the new automation by being in the control room at key times to adjust output tracking values, ratios, feedforward dynamic compensation, and PID tuning.

Watch-outs: Erratic measurements at low flows and excessive valve stiction near the seat may prevent the PID from controlling at low flows. Analyzers may give unusable values or provide no new values. Use logic to screen out unreasonable analyzer values and to turn off integral action, or use an enhanced PID (Tip #100) to avoid ramping of the PID output when an analyzer fails to update.

Exceptions: If the signal is too erratic or unreliable, closed loop control is not possible unless an inferential measurement can be developed and corrected online by an at-line or off-line direct measurement.

Insight: Close loop control can provide a response that accounts for unknowns and is more consistent, timely, and amenable to improvement and optimization than the best process engineer or operator can provide.

Rule of Thumb: Use PID control to eliminate manual actions and scheduled actions whenever a usable PV measurement is available.

Tip #70: Minimize Deadtime[1,2]

Greg McMillan (Technical)

When I was four years old and sitting on my daddy's knee, he said, "Son, I have just one word for you: deadtime." I did not understand the significance of this word of wisdom for decades. The math in my control theory classes mostly served as a distraction from the essential truth: if the deadtime is zero, the controller gain can be infinite and the reset time zero. The controller has enough muscle for an instantaneous response. The errors from disturbances can be zero.

Without deadtime, I would be out of a job. The importance of deadtime particularly hit home in the pH control of systems with steep titration curves, where the slope and hence the process gain can change by a factor of 10 for each pH unit deviation from setpoint. Reducing deadtime reduces the excursion on the titration curve, minimizing the nonlinearity seen by the pH control loop.

You can get a feel for deadtime by drinking Hurricanes on Bourbon Street. The time from your first drink to a feeling of being more of a party person than an engineer is deadtime. If you drink too many Hurricanes in this deadtime, you may be out of control.

The key role of deadtime in tuning and loop performance is largely missing in the control literature. Fortunately, I found Greg Shinskey as a guiding light. Shinskey's articles and books listed in Appendix A and B offered the best knowledge of process relationships and dynamics, focusing particularly on deadtime as the culprit. Without this essential understanding, you are vulnerable to a lot of misconceptions. For example, I saw an academic paper where there was no deadtime in the simulation. The author was proudly showing how his special tuning method reduced the control errors. He did not realize that a tuning method was not even required. He did not understand that it was possible to continually increase the controller gain and improve loop performance.

Concept: Deadtime delays the loop's identifying a change and making a correction. The loop deadtime is the total time delay for a complete loop around the block diagram from any starting point; that is, the sum of actuation (e.g., control valve), correction (e.g., injection), process, recognition (measurement), and execution (controller) delays and the equivalent deadtime from lags (time constants) that are smaller than the largest time constant in the loop. For unmeasured disturbances and controllers tuned for maximum disturbance rejection, the peak and integrated errors are proportional to the deadtime and deadtime squared, respectively (Tip #71). Process, mechanical, and control system designs should minimize the total loop deadtime to fundamentally increase the ability of the control loop to do its job. The ultimate period (inverse of the natural frequency in cycles per sec) is proportional to the total loop deadtime. For most processes the maximum controller gain and minimum reset time are inversely and directly, respectively, proportional to the ultimate period and thus the deadtime.

Details: Develop an understanding of the sources of deadtime. Process delays (e.g., mixing and transportation delays) create a continuous train of delayed values. Digital delays cause a discontinuous update at discrete intervals. The equivalent deadtime from digital delays is ½ the cycle time plus the latency (delay from the start of cycle time to the report of result). For most digital devices, the latency is negligible. The deadtime from wireless devices is ½ of the default update rate for changes that do not exceed the trigger level. The deadtime from a PID controller in a distributed control system (DCS) is about ½ of the module execution time. The deadtime from an analyzer where the result is at the end of the analyzer cycle time is 1½ times the cycle time plus the sample transportation delay. An enhanced PID developed for wireless can enable more aggressive tuning when deadtime from the digital device or analyzer is larger than the process time constant. The deadtime from actuator and positioner sensitivity limits, from valve backlash and stick-slip, and from digital signal quantization is the deadband, resolution, and threshold sensitivity divided by the rate of change of the input to the respective automation system component. The deadtime from automation system time constants (e.g., sensor lags, transmitter damping, and signal filter times) that are small compared to the rate of process change can be approximately taken as equivalent deadtime. For large loop deadtimes, feedforward control is advisable for measurable large and/or fast disturbances. When the deadtime becomes much greater than the open loop time we have a case of deadtime dominance. Tuning methods break down and peak errors for step disturbances are as big as if there were no feedback control. For a list of solutions in this case, see the July 19, 2012 Control Talk blog "Deadtime Dominance Does Not Have to Be Deadly."

Watch-Outs: Field and simulation tests or imagined scenarios where the disturbance always occurs just before the input of the PV will not show the increase in deadtime caused by a digital device. Such tests or scenarios can lead to the erroneous conclusion that digital delays do not increase the ultimate period or deadtime. Disturbances can arrive at any point in a digital device's cycle time and should be thought on average as arriving halfway through the cycle time. Tests or scenarios where the input step change is larger than the deadband, resolution, and threshold sensitivity limits will not show additional deadtime from these limits. Deadtime compensators cannot eliminate the effect of deadtime on the ultimate limit. The more aggressive PID tuning possible for deadtime compensators is dependent on an exceptionally accurate quantification of deadtime with an underestimate less than 25% and an overestimate less than 10%.

Exceptions: The equivalent deadtime from small time constants has a factor (fraction of the time constant converted to deadtime) that increases as the ratio of the smallest to largest time constant in the loop decreases.

Insight: Deadtime is the ultimate limit to how well a loop can correct for unmeasured disturbances and how aggressive you can tune the controller's response to be.

Rule of Thumb: Minimize the major sources of deadtime and consider the use of enhanced PID and feedforward control for large irreducible deadtimes.

Tip #71: Achieve Best Disturbance Rejection and Setpoint Response[1,2,3,4,5,6,7,8,9,10]

Greg McMillan (Technical)

If you don't want to get into the technical jibber-jabber, just skip to the Insight and the Rule of Thumb. Be content without the content. If you don't want to be distracted by reality, get your kicks in the math. Design new feedback control algorithms, despite the fact that PID has been proven to be time-optimal for unmeasured disturbances since 1976, as detailed in the paper by Alan Bohl and Tomas McAvoy "Linear Feedback vs. Time Optimal Control. II - The Regulator Problems"[10]. Focus on setpoint response and don't be distracted by the reality that loops on continuous processes rarely have setpoint changes, special logic can be developed because a setpoint change is a quantitatively known event, and batch loops have to correct for disturbances and changes in demand. If you have an upset, put it on the process output instead of the process input so it bypasses the process and is like measurement noise. On the other hand, if you have control loops in your plant that affect product quality, onstream time, or process efficiency, please read this tip.

While working on difficult loops such as incinerator and phosphorus furnace pressure control and compressor surge, where disturbances could cause a trip within a second, and in pH control where disturbances could drive the pH to the limits of the scale and the operators' patience, I was focused on how the controller could react to minimize peak error. In the literature there is little written on peak error, but in a gem of a book by Peter Harriott, *Process Control*, I found a simple equation for the peak error[9].

I moved on to figuring out how to minimize energy use and the amount of off-spec material. Here, I recognized that the integrated error was important, where plus and minus areas of the error on a trend chart cancel out like the errors do in large equipment. I found a simple equation in Greg Shinskey's books listed in Appendix B that showed that the integrated error (IE) was proportional to the reset time and inversely proportional to the controller gain. However, the control literature was totally focused on integrated absolute error (IAE). I resolved the discrepancy by realizing that if the controller was tuned for a non-oscillatory response, the IE and IAE were identical.

I next sought to minimize the time to reach setpoint in automated start-ups and in batch operations. For bioreactors, I found that pH and temperature overshoot were critical and time was not. Furthermore, the disturbances from changes in cell metabolism were so slow that disturbance rejection was unimportant.

Concept: Loop performance objectives should fundamentally address the need to minimize the process variable (PV) response to disturbances and to maximize the PV response to new setpoints (SP). Disturbance objectives are minimization of peak error and integrated absolute error (IAE). Setpoint objectives are minimization of overshoot, settling time, and rise time (time to reach setpoint). The speed of PID tuning sets the practical limit on loop performance for these objectives. Fast (aggressive) tuning reduces peak error, IAE, and rise time. A PID can be tuned faster if the deadtime and the PV ramp rate for a given change in PID output are decreased. Minimization of overshoot, traditionally achieved by slow PID tuning, can now be achieved by using key PID features without sacrificing other loop performance objectives.

Details: All processes have unmeasured disturbances. Minimize peak error to prevent undesirable reactions, safety instrumented system (SIS) or relief activation, and exceeding environmental limits. Minimize the IAE to reduce the quantity of off-spec produced and the quantity of utilities and raw material used. Minimize both peak error and IAE by maximizing gain and minimizing reset time. Maximizing gain is more important for peak error. Overshoot can cause many of the same problems as peak error. Rise time is important for minimizing cycle time in batch processes and minimizing start-up and grade transition time in continuous processes. Minimize overshoot and rise time by increasing reset time and gain, respectively. Add a setpoint filter time equal to reset time to prevent overshoot in a PID with fast tuning. Minimize settling time by minimizing overshoot and rise time. Add logic for smart sequenced positioning of final control elements and setpoint feedforward to further enhance the setpoint response. In the sequencing of controller outputs, position and hold the output at the appropriate output limit until the rate of change of the PV multiplied by the deadtime is near the setpoint. At this point, position and hold the output at a final resting value for one loop deadtime (see Tip #91 for more details).

Watch-outs: Fast (aggressive) tuning decreases the robustness of the controller (its ability to retain a smooth response for increases in deadtime and PV ramp rate for a given output change). External-reset feedback (dynamic reset limit), described in Appendix E, must be used to prevent a burst of oscillations from fast tuning (high controller gain and/or low reset time), causing the PID output to change faster than a final control element or secondary loop can respond for large disturbances or setpoint changes. The use of a setpoint filter in a secondary loop may degrade the ability of the primary loop to reject disturbances.

Exceptions: For processes with exceptionally slow disturbances (e.g., cell culture changes in bioreactors), the peak error and integrated error are inconsequential, even for slow tuning. For batch operations with long cycle times or continuous processes with long start up times that are sensitive to operating point, minimizing overshoot is more important than minimizing rise time (e.g., bioreactor temperature and pH).

Insight: Loop performance objectives can be achieved by maximizing controller gain and minimizing deadtime, reset time, and ramp rate.

Rule of Thumb: Use disturbance rejection tuning, external-reset feedback, and a setpoint filter in your PID controller to achieve loop performance objectives.

Tip #72: Achieve Best Loop Coordination and Optimization[4,5,6,7,8]

Greg McMillan (Technical)

As a modeling and control specialist in the Engineering Technology department of a large chemical company, I was called in to solve difficult and persistent problems that had plagued the production units. I focused mostly on important individual loops in key unit operations, such as gas pressure loops for furnaces and headers, temperature loops for reactors and kilns, surge loops for compressors, level loops for boiler drums and surge tanks, pH loops for all types of equipment and streams, dissolved oxygen loops for bioreactors, polymer pressure loops for extruders, and thickness control loops for sheets.

Except for headers, kilns, plug flow reactors, gas reactors, and sheets, problems with interacting loops were minimal. I could focus on individual loop performance and on transferring as much variability as possible from a key process variable (PV) to its manipulated variables. Even for unit operations with interacting loops, a simple half decoupler with the feedforward of one loop output to another loop output would usually suffice. In distillation, my associate Tolliver did not need to do much in the way of decoupling because there was usually only one temperature loop. Two point composition control, that is, control of top and bottom composition by temperature, was rarely needed.

I generally did not have to worry about coordination or the movement of a manipulated variable affecting another loop. I could go for the gusto. For optimization, I simply used valve position controllers (VPC) to maximize feed by pushing coolant, vent, and feed valves to a maximum throttle position (Tip #97). I also used VPC to maximize the use of waste fuels and reagents by pushing an expensive fuel or reagent valve with good process dynamics to a minimum throttle position.

In petroleum refining, the gas furnaces and catalytic reactors have interactions and don't have much back mixing reducing the advantage of the PID. There is a greater need for plant-wide optimization. Because of incredibly large plant capacities, a fraction of a percent improvement in efficiency or production rate translates to huge increase in profitability in an industry where margins are squeezed. Consequently, MPC, linear programs (LP), and even real-time optimization (RTO) are used. I know of one refinery where all control systems were moved from PID to MPC because of the in-house MPC expertise.

I have used MPC with success but have returned to PID because existing and new PID features can be used in more innovative ways than expected to meet process objectives. I have to acknowledge I am the product of Shinskey's books (see Appendix B). I like to tackle the process on a unit operation basis.

Concept: Loop coordination objectives are used to ensure nearly identical SP responses and to decrease loop interactions. A primary optimization objective is to achieve a more efficient or productive operating point without upsetting the process. The effective use of key PID features can achieve these objectives[2].

Details: There are three major loop coordination objectives plus an optimization objective. The first coordination objective is to ensure the simultaneous response of flow inputs to a unit operation (e.g., ratio flow control and blending). The second is to all secondary loops to have a uniform desired secondary PV response to a setpoint change in cascade control or model predictive control (MPC). This secondary PV response consistency is particularly important for parallel unit operations, so that differences in the primary controller tuning and MPC model are indicative of discrepancies in identical unit operations. The third coordination objective is to slow down the transfer of variability from the PV to the PID output to reduce interactions. Coordination has traditionally been achieved by increasing reset time and decreasing gain. However, this tuning sacrifices disturbance rejection and rise time (Tip #71). Optimization has been traditionally achieved by integral-only valve position controllers that are difficult to tune. Today, there are key PID features (e.g., external-reset feedback) to provide an easier and more effective means of coordination and optimization. Use external-reset feedback with directional setpoint rate limits or a setpoint filter to provide the desired speed of response with disturbance tuning. For optimization loops, use a slow setpoint rate limit to provide a gradual approach to the optimum and a fast setpoint rate limit to provide fast correction for disturbances and abnormal operation. For surge loops, use directional rate limits for the analog output setpoint to provide a fast opening and slow closing surge valve. For valve position controllers that maximize coolant temperature setpoints or reactor feed setpoints to push coolant valves to the largest safe throttle position, use slow increasing rate limits and fast decreasing rate limits on the process controller setpoint being maximized (Tip #97).

Watch-Outs: Setpoint rate limits on analog outputs (AO) will confuse Maintenance and Operations when they are manually stroking a valve. AO setpoint rate limits should be turned off when in manual. Controllers without the positive feedback implementation of integral action may not have external-reset feedback.

Exceptions: Optimization must take a back seat to the correction of large, fast disturbances. To prevent equipment damage and an environmental violation, an open loop back up may be needed (Tip #91).

Insight: The response of control loops or final control elements can be coordinated to work together to meet process objectives by the use of setpoint filters and velocity limits and external-reset feedback.

Rule of Thumb: Set a setpoint filter time that the slowest coordinated loop can achieve, slow down the AO block setpoint rate limit in both directions to minimize interaction, and set the AO block setpoint rate limit slow in the optimization direction and fast in the protection direction.

Tip #73: Use Coriolis Meters for Mass Flow and Concentration Control[9,10,11]

Greg McMillan (Technical)

In the 1980s and 1990s, Coriolis meters were expensive and were primarily relegated to critical measurements of small liquid streams. The meters were susceptible to vibration and errors from bubbles or solids. Since then the price and capability have improved. Coriolis meters can now be used on gas streams and can measure the percent bubbles and percent solids. Density measurement has always been a strong point, with a precision out to the 4th decimal place for grams per milliliter, and there is no drift and very little maintenance required.

Process gains and online metrics (Tip #61) are a function of mass flow ratios. We have gotten used to working in volumetric units because volumetric flowmeters are so prevalent. However, you do not hear of a process engineer doing a volume balance. Process analysis and modeling require a material balance for the conservation of mass. Material balances often do not close due to errors in measurements made with volumetric meters. Even when they are corrected for pressure and temperature, mass flow computations based on volumetric flow measurements assume a known and constant composition, which may not be the case. Coriolis is the only means of true mass flow measurement.

Coriolis meters have been commonly used in the front end operations of specialty chemical, food and beverage, and pharmaceutical plants. The front end operations, such as reactions, set product quality and yield. Accurate mass flow ratios enable maintaining stoichiometric ratios. Coriolis meters are now moving downstream as their precision and rangeability advantages are recognized. Coriolis meters are replacing load cells for inventory management and are increasingly being used in recovery, blending, and custody transfer. Materials of construction have expanded to include titanium. The principal remaining reason not to use a Coriolis meter in a process stream is cost for larger pipelines. However, if the cost analysis included the process benefits from better process knowledge and turndown and the elimination of downtime and maintenance, Coriolis meters would be justified for most process streams.

There is often a misunderstanding about the performance of Coriolis meters because of the different models and vintages. The newest and most capable meters, with technological advancements in sensor design and signal processing, offer an order of magnitude improvement in resolution over earlier versions. A turbine meter may provide slightly better resolution, but only for a perfect velocity profile and for fixed composition and viscosity. In addition, a turbine meter is a mechanical device that requires maintenance and has only a 15:1 rangeability, whereas a Coriolis meter has a 200:1 rangeability.

Concept: Blending, crystallization, distillation, evaporation, and reactions benefit from material balance analysis and control from online measurement of mass flow rates and the composition of the important liquid streams. The density measurement capability of the Coriolis meter can be used as an inferential measurement of component concentrations and solids for two components or a single component, two phase mixture. The flow rate and concentration measurements can be used for intelligent feedforward signals (Tip #92), process modeling, online data analytics, online process metrics (Tip #61), and live process flow diagrams on operator screens (Tip #101).

Details: The best U-tube Coriolis meters maintain an extraordinary accuracy of 0.05% and a repeatability and resolution of 0.02% of rate for liquids, independent of concentration, viscosity, temperature, time, and upstream/downstream straight pipe runs. There is no drift. Once properly calibrated and installed in an application, there is no maintenance required. The density measurement is accurate to 0.0002 gm/cm^3. The flow measurement rangeability is 200:1. Advanced meter software is capable of measuring bubble and solids concentrations. Sizes range from 2 mm to 300 mm (0.1 to 12 inch). Use the most accurate Coriolis meters for all important process streams.

Straight tube Coriolis meters, used to prevent solids accumulation or erosion, have an error and repeatability about five times larger. Error and repeatability in meters used for gas flows are about 10 times larger than for liquid flow.

Watch-outs: Misalignment of flanges can put torque on the tube, creating significant error. Vibration and crosstalk from adjacent meters can cause noise. Solids can accumulate on the bottom, and erosion can occur at the turns of the U-tube when streams contain abrasive particles. Bubbles can accumulate in the meter when streams contain gases or vapors. For streams that contain solids, it is better to use straight tube meters installed vertically for self-cleaning. For streams containing gases or vapors, installation must prevent trapping bubbles or causing flashing or cavitation in the meter. For gas flows, low density due to low molecular weight and/or high temperatures will reduce accuracy, sensitivity, and repeatability. Less expensive and older U-tube models and straight tube designs offer much less accuracy and precision.

Exceptions: The cost may be prohibitive in line sizes above 6″ unless the reduction in maintenance and the benefits of tighter material balance analysis and control provide an acceptable rate of return on capital investment. Particles and coatings may clog the meter in line sizes below 1/2″. Maximum temperature is 350 °C and materials of construction are more limited than for other flowmeters. Coriolis meters are not often used in oil and gas platforms and pipelines, refineries, and petrochemical plants because process knowledge is well-established and the line sizes are large.

Insight: Coriolis meters require the least maintenance and offer the best long-term accuracy of mass flow and density measurement, independent of piping and process operating conditions for liquid and gas streams, which enables better advanced control and process analysis.

Rule of Thumb: Use Coriolis meters for measuring liquid flows in batch, fed-batch, and continuous operations where tight flow ratio and composition control, accurate process modeling, and high turndown are beneficial.

Tip #74: Eliminate Split Range Oscillations

Greg McMillan (Technical)

Just as split personalities cause mental instability, split ranged loops cause process control instability. Actually, the situation is worse for split ranged loops because valve stiction is particularly bad near the closed position, and the transition between fluids (e.g., between water and steam) may involve changes in phase besides changes in dynamics.

If you have a split ranged loop, I bet you 25 cents that the loop oscillates at times across the split range point. If oscillations never occur when operating at the split range point, please mail me a copy of the trend chart, along with a self-addressed, stamped envelope, and I will mail you the quarter.

Stiction causes a limit cycle in a loop with integrating action at one or more points in the process or in the controller. The introduction of a deadband into the split range point in an attempt to stop unnecessary crossing back and forth of the split range point also causes a limit cycle if there is integration in two or more points in the process or control system. Batch temperature, pH and composition control, vessel pressure control, and vessel level control all have integrating action in the process. Thus, the addition of one PID loop causes limit cycles from deadband. The use of cascade control, where both the primary and the secondary loops have integral action, will cause a limit cycle regardless of process type. Deadband also appears in control valve backlash and in variable speed drive setup.

We think about the difference in process gain for a change in phase in the fluid being manipulated. We may not think of the difference in process deadtime and time constant for the same change. When there is a change in phase, there is also a significant change in process dynamics. For example, as a reactor temperature controller switches from steam to cooling water, or a bioreactor pH controller switches from carbon dioxide (gas) to sodium bicarbonate (solid), there is a big change in the process time constant. In addition, there is an undetermined residual effect of one phase. For reactor jackets, switching from cooling water to steam can result in pockets of water and droplets in the steam. Switching from steam to water can result in pockets of vapor and bubbles in the water. The result is cold and hot spots and an erratic response in the transition.

In chemical reactors, there is often a time when there is no significant demand for heating or cooling. In bioreactors there is often no significant demand for an acid or base. The best thing for the control system to do is nothing but because of integrating action, limit cycling will occur.

Needless to say, most loops oscillate at the split ranged point, decreasing process efficiency and in some cases decreasing product quality. For bioreactors, the unnecessary addition of sodium bicarbonate increases cell osmolality (internal cell pressure), causing cell rupture and death. A batch worth several million dollars can be lost due to high osmolality from oscillations at the split range point.

Concept: Most split ranged loops oscillate across the split range point. Stopping these oscillations requires a comprehensive approach, addressing changes needed in the process, control system, and final control elements. Missing just one of the many sources will result in continuing (but usually less severe) oscillations.

Details: To avoid the discontinuities and unpredictable behavior from the changes in phase between using steam and cooling water, use steam injection so the transition between heating and cooling is a transition between hot water and cold water. To reduce the effects of stiction and backlash at the split range point, add a small sliding stem valve with a diaphragm actuator and digital positioner for fine adjustment at the split range point. Even if this does not stop the oscillations, the consequences in terms of process efficiency and product quality will be less severe. The fine adjustment valve must be kept from going wide open by the manipulation of a large valve by a valve position controller (Tip #97). This strategy can provide greater precision at all operating points. Do not add deadband in the split range control configuration or in a variable speed drive setup. Use an auto-tuner or adaptive tuner to identify the different process dynamics of split ranged valves or secondary loops and schedule the tuning based on the primary controller output. Use external-reset feedback and directional setpoint rate limits on an analog output block for split range valves and on PID setpoints for secondary loops to suppress movement into the split range point. To stop limit cycles use an enhanced PID developed for wireless with a threshold sensitivity setting to ignore noise and insignificant changes.

Watch-outs: The split ranging of a small valve and a large valve will create another point of oscillations. The valve position controller eliminates the addition of a split range point but will interact with the process controller unless an integral-only structure or an enhanced PID is used. For exothermic reactors, transitions from heating to cooling *should not* be suppressed and valve position controllers *must* not slow down the need for more cooling to prevent a runaway. Valves with very low leakage specs tend to be on-off valves. Do not use on-off valves for any throttling service (Tip #84). For tight shutoff and positive isolation use an on-off valve that is coordinated with the opening and closing of the small valve.

Exceptions: For valves that have difficulty opening after being closed for a long time due to temperature or phases, the periodic opening of the valves caused by split range oscillations may help prevent them from seizing.

Insight: Split range oscillations occur due to process nonlinearities and discontinuities, control valve deadband and stiction, and controller integral action and overreaction.

Rule of Thumb: Use precise valves, scheduled tuning, external-reset feedback, directional rate setpoint limits, and an enhanced PID to stop split range oscillations.

Tip #75: Use Wireless Transmitters for Diagnostics, Improvements, and Metrics[11,24,31]

Greg McMillan (Technical)

In the 1970s, there was a concerted effort at the chemical company I worked for to provide extra process measurement connections throughout the plants for pressure and temperature measurement. These connections could be used as needed to help identify and solve unforeseen problems. The installation of a wired transmitter on a temporary basis was not practical because of the cost to install conduit and wiring that is safe, reliable, and noise free. Consequently, the company developed standards for pressure and temperature gauges that had scale ranges and materials of construction to meet most application requirements. Unfortunately, only the operator in the field standing in front of the gauge could see the indication and the resolution was poor. I still have flashbacks of how a kind old instrument engineer spent the remaining years of his career figuring out how to compromise scale ranges.

With modern wireless smart transmitters, we could have obtained this information far more easily and accurately. Smart transmitters enable ranges to be readily configured without much of a difference in accuracy. If a gateway is set up and other wireless transmitters are already being used in the plant, the commissioning of a new transmitter may just be a matter of minutes. Line of sight is usually not required because other wireless devices can be used as necessary to find a way to the gateway. Channel hopping eliminates noise concerns. Encryption and invitation-only communication eliminate security concerns. The measurements can be readily indicated, historized, trended, and analyzed by operators and engineers. Wireless offers portability, analysis by asset management systems, historization, and computations for metrics, with the accuracy and diagnostics afforded by smart instrumentation.

Pressure is the driving force for material flow, and temperature is the driving force for heat flow. Flow disturbances start out as a pressure change and heat transfer disturbances start out as a flow or temperature change. The fouling of a heat transfer exchanger can show up as a change in the temperature or pressure difference between the tube side inlet and outlet streams.

Heat transfer rates and overall heat transfer coefficients can be computed if the heat transfer area and flow are fixed or measured. Recirculation flows are often relatively constant. When a heat exchanger in a recirculation line is used for vessel temperature control, cascade control system may already provide the necessary measurements. The cooling heat transfer rate is simply proportional to the primary loop temperature minus the secondary loop temperature; heat exchanger process inlet minus outlet temperature.

As I started to develop the use conductivity and pH for CO_2 reduction at the University of Texas Pickle Research Campus, I realized there was a whole new opportunity for wireless. pH electrodes were failing and the relationships to CO_2 load and solvent concentration needed to be quantified. The best sensor type and location needed to be found. Lab test results had to be historized and analyzed.

Concept: Wireless transmitters can be easily stocked, configured, and installed to explore and demonstrate diagnostics, metrics (Tip #61), and improvements, such as inferential measurements, new sensor technologies, new strategies, and better sensor locations.

Details: Use wireless pressure measurements to track down problems such as field regulator problems, screen and filter plugging, heat transfer surface fouling, and insufficient valve pressure drop as a percent of total system pressure drop. Use wireless pressure transmitters to provide the pressure profile for a piping system and enable the computation of the installed characteristic of a control valve. Use wireless temperature transmitters to provide an inferential measurement of heat transfer rate and overall heat transfer coefficient. Heat transfer rate measurements can enable the optimization of the scheduling of batches for energy conservation and the tradeoff between yield and production rate in batch reactors. Use wireless temperature transmitters to provide a temperature profile in a utility system for pinch analysis to find the location of the limitations in the utility system for better process efficiency and capacity. Use wireless temperature transmitters on distillation columns to find the tray with the largest and most symmetrical change in temperature for a change in reflux to distillate ratio or steam to distillate ratio. Use wireless acoustic and temperature transmitters to monitor steam traps to detect when traps are stuck open or closed. Traps stuck closed can back up condensate into the heat transfer volumes of process equipment, severely disrupting process temperature control systems. Traps stuck open will blow through steam, causing a loss of energy and higher condensate system pressure, which can cause other steam traps to shut, backing up condensate into other process equipment. Traps that are alternately stuck open and closed cause confusing and erratic plant operation. Use wireless pH transmitters in the lab in worst-case process samples to determine electrode performance, life expectancy, and warning signs of electrode failure for predictive maintenance. Use different electrode types in the lab sample to find the best sensor technology. Use wireless pH transmitters in the field to find the measurement location with the best mixing, least noise, and least deadtime (Tip #67). Use wireless mass flow computers and annubars to provide flow measurements where needed to develop and demonstrate online diagnostics and metrics. Consider a system of plant spares where wireless transmitters can be loaned to enable the rapid transition from idea to a working prototype in the field.

Watch-outs: Spare process connections may not have been provided in the design of the plant. Batteries should be monitored. The default update rate for wireless pressure measurements may not be fast enough to track down disturbances.

Exceptions: Wireless transmitters that use a lot of electrical power are not available. Wireless transmitters are too slow for surge detection and control, compressor control, polymer pressure control, and furnace pressure control.

Insight: Wireless transmitters eliminate the cost and time hurdles for moving innovations from research, engineering, and operations into action, enabling the rapid demonstration of benefits and inspiring creativity.

Rule of Thumb: Use wireless transmitters to rapidly prototype, demonstrate, analyze, and justify installations of online diagnostics, metrics, and improvements.

Tip #76: Use Checklists to Cover All the Bases

Greg McMillan (Technical)

The information needed to successfully select and install instruments is spread out in articles, books, handbooks, papers, and vendor catalogs and in knowledgeable individuals who do not have the time or incentive to publish.

Articles and papers tend to focus on success stories, with few details. The articles are designed more to sell than to help implement the automation system component. Even if the article or paper lists a user as the author or coauthor, it was probably written by a marketing or sales engineer of the supplier. This is not to say the article or paper does not provide valuable technical content or is misleading. These people know how to present information in a readable and understandable manner.

Books and handbooks tend to give the principle of operation or theory, which is great, but you need to move on to all of the practical considerations for a successful application. Information and guidance for instrument and equipment selection, configuration, and installation is often missing in the literature. *Flowmeters - A Comprehensive Survey and Guide to Selection* is an example of an ISA book on the right path. I tried to provide guidance in my book *The Essentials of Modern Measurements and Final Elements*. However, the reader has hundreds of pages to read. Also, despite the author's best efforts, some things may be missing because they were not known to be important at the time. I decided something better needed to be done. I concluded what would be helpful would be succinct insights, rules of thumb, and checklists. In the future, all of my books will have these features. With this in mind, I have included all of my checklists to date in Appendix C. The checklists ask questions to make sure you are covering all the bases. You can pose these questions to your suppliers of automation products and services or use them in searches of online publications. In the long run, you can start to realize better how to address all aspects in the selection, installation, and use of automation system components.

I personally believe the best way to present essential information in our profession involves checklists in addition to concise statements of concepts, details, watch-outs, exceptions, insights, rules of thumb, and resources exemplified in this book. For presentations, the statements should be bullet points.

Concept: The information needed to take advantage of the best technology and to address all of the considerations of an application is often difficult to find. Engineers don't have time to dig through dozens of sources and hundreds of pages for each automation system component. Search techniques don't help much if you don't even know what questions to ask. Suppliers may not know your application well enough to help you make the best choice. Checklists can succinctly list the right questions to ask.

Details: Use the checklists in Appendix C to "zero in" on what you need to know about control valves, differential pressure and pressure transmitters, inline flowmeters, temperature sensors and transmitters, pH electrodes and transmitters, PID controllers, radar level devices, trend charts, and variable speed drives. Use the key questions for control valves to avoid specifying a valve with excessive deadband, poor resolution and threshold sensitivity, and insufficient rangeability. Use the key questions for inline flowmeters to minimize drift and noise and extend rangeability. Use the key questions for temperature sensors to minimize drift and conduction error and improve precision. Use the key questions for pH electrodes and transmitters to minimize drift and improve response and life expectancy. Use the key questions for PID controllers as to external-reset feedback, setpoint filter, setpoint rate limits, and wireless enhancement to achieve process objectives. Use the key questions for radar level to minimize erratic signals from swirling, turbulence, and low dielectric constant. Use the key questions on trend chart compression, variables to plot, and time scale to help analyze loop and process performance. Use the key questions on variable speed drives to minimize loss of rangeability from overheating, slip, and high static heads.

Watch-outs: The checklists cannot cover all possible scenarios and the effects of harsh process conditions. For example, high temperatures can degrade sensor performance and cause premature sensor failure even though the temperature is below the supplier's specified high limit. This deterioration is particularly significant for a resistance temperature detector (RTD) and a pH electrode. High temperatures may require remote mounting of the transmitter. Wetted materials of construction including gaskets and O-rings must meet the worst-case corrosion conditions including abnormal operations.

Exceptions: There are exceptions to every rule and every checklist. Plant practices and procedures important for a safe and reliable installation and operation are not included in the checklists. Except for pH, maintenance considerations are not covered.

Insight: Checklists can provide the questions to ask yourself and your resources.

Rule of Thumb: Make sure each question on the checklist is answered or is determined to be inapplicable.

"Checklists ask questions to make sure you are covering all the bases."

Tip #77: Increase the PID Gain and Reset Time for Vessel and Column Control[1,2]

Greg McMillan (Technical)

I have found over and over again that the simplest and best thing you can do for level, pressure, pH, and temperature control on vessels and columns is to simply increase the reset time by a factor of ten. It is truly amazing how many of these loops are in a reset cycle. Current practices for operator interfaces and human nature are big reasons for loops having too much integral action.

Operator screens typically have lots of equipment and piping graphics but show digital values for operating conditions. The PID controller faceplate shows the setpoint (SP), process variable (PV), and output (OUT) as bars besides the digital values. The operator looking at these displays has no sense of direction. Several times I have gotten urgent phone calls that something was wrong with the PID controller. In one instance, the operator was looking at the faceplate for a temperature loop on a reactor, where cooling occurred below and heating occurred above the split range point. The temperature was below setpoint by a degree or two and the operator, who was intently watching the faceplate, said the PID needed to be fixed because the cooling water valve was open instead of the steam valve. I have shown a PID with similar conditions in presentations and asked which valve should be open. I haven't had one person answer that the cooling water valve should be open. However, if I show a trend chart of the PV trajectory, I get more correct answers.

Integral action will do what a human would do. Integral action just knows the temperature is low and thinks heat should be added. In this case, the contribution from the integral mode is to continually ramp up to add more heat. The contribution from the proportional mode is to decrease to add cooling. People looking at the faceplate naturally decrease the reset time to get what they think is the proper action. People have no sense of deadtime and little sense of trajectory unless the trend chart is set up to display them (for example, per the Checklist in Appendix C). For more on this problem, see the modeling and control March 4, 2011 blog "Are we Misleading our Operators?"

Processes with a slow, gradual response by virtue of a large, back-mixed volume afford the opportunity for a higher PID gain. Vessels have back mixing by virtue of agitation and recirculation. Trays in columns have back mixing from the turbulence of vapor flow going up and reflux flow coming down. The response of each volume is mostly characterized by a process time constant for continuous operations. Vessels can have a process time constant of an hour. Large columns can have a process time constant of five hours. Equipment with large process time constants has a slow, gradual process response. For level and batch operations, volumes have an integrating response that is extremely slow, perhaps 0.0001% per sec. These processes benefit from more proportional action to provide enough muscle to make the process move faster. In most cases, the output should be driven past its final resting value to get the PV to setpoint faster.

Concept: Vessels and columns have a gradual response that requires muscle, with preemptive action based on the approach to setpoint. The controller gain provides this action through the proportional mode. Integral action has no sense of the approach to setpoint and will cause overshoot. Most temperature loops on vessels and columns have a gain that is too low and a reset time that is much too low. Temperature noise is normally negligible due to the slow gradual response of the volume.

Details: Ballpark estimates of useful gain and reset times for temperature loops on large, well-agitated liquid reactors are 25 and 10 minutes, respectively. Ballpark estimates of useful controller gains and reset times for temperature loops on columns are 2.5 and 100 minutes, respectively. Use a near-integrator tuning with Lambda times approaching the deadtime to get these tuning settings. Pressure loops on this equipment may have a controller gain that is about the same as the temperature controller gain. The reset time for pressure control is usually much smaller than the reset time for temperature control, especially for columns, because the vapor response is much faster than the liquid composition response. Noise free level loops on this equipment may have a controller gain five times larger than the temperature controller gain. For horizontal reflux drums, the level controller gain can be even higher. The actual controller gains applied depend upon flow and temperature measurement scale ranges, valve sizes, and process gains. For pH loops on well-mixed vessels, the reset time is comparable to the temperature controller reset time. The controller gain depends heavily upon the slope of the titration curve. The pH controller gain can be as high as 5 for weak acids and weak bases. For strong acids and bases, the controller could be a factor of 100 lower.

Watch-outs: The controller gain is proportional to the time constant time to deadtime ratio but is also inversely proportional to the open loop gain. The open loop gain is the product of the manipulated variable gain, process gain, and measurement gain. Consequently, the controller gain depends upon valve sizes, process relationships, and measurement scales. A reasonable reset time is about four times the deadtime. Runaway reactors may require a reset time that is a factor of ten larger. For temperature loops on vessels, the deadtime depends upon the turnover time from mixing, thermal lags, and sensor lags. For columns, the deadtime depends on the number of stages and the volume per stage. See the July 2012 Control Talk blog "The ABCs of Tuning" for the inside story on tuning.

Exceptions: Vessels with poor mixing or narrowed measurement scale ranges have lower controller gains. Column temperature loops that manipulate steam flow will have much smaller reset times.

Insight: Temperature loops on mixed volumes, gas pressure loops, and level loops need much larger than expected controller gains and reset times.

Rule of Thumb: See if increasing the reset time by a factor of ten for loops on vessels and columns will reduce overshoot and oscillations.

Tip #78: Decrease the PID Gain and Reset Time for Pipeline and Inline Control[1,2]

Greg McMillan (Technical)

I quickly started to appreciate pipeline and inline control from working on inline pH control where a static mixer (a baffled piece of pipe) was the only source of mixing. Static mixer manufacturers touted radial mixing, but no back mixing, that is, there was only plug flow. No back mixing meant that the components at the static mixer inlet all arrived at the outlet at the same time. All of the components spent the same time residing in the static mixer. This tight residence time distribution is beneficial when static mixers are used as plug flow reactors to provide consistent conversion; reaction rate multiplied by residence time. However, for pH control that had an effectively instantaneous neutralization rate, I was looking for some back mixing so that a portion of the residence time (volume divided by volumetric flow rate) was a process time constant.

I appreciated one supplier proudly quantifying the lack of back mixing; it simplified my decision not to buy the advertised static mixer. The supplier's sales engineers could not understand what I needed for pH control. This failure to have a meaningful dialog on the effects of equipment design on process control comes down to a lack of recognition that a process time constant is good and process deadtime is bad for tight process control.

Pipeline and inline control loops have little back mixing. The residence time is transportation delay and hence pure deadtime. The only time constants in the loop are from the sensor, transmitter damping, and signal filters. Flow, blend concentration, conveyor or feeder load, desuperheater temperature, extruder temperature, liquid and polymer pressure, jacket inlet temperature, sheet thickness, pulp stock consistency, and static mixer pH loops (among others) effectively have no process time constant. Heat exchangers, inline reactors, and fluidized bed reactors essentially have plug flow but have a series of process time constants in the temperature response from the thermal lags associated with the heat transfer surfaces.

The lack of process time constant means there is no filtering of process noise by the process and the process response is not gradual for a step change in a process input. Consequently, manual actions, sequential actions, on-off control, and oscillations in loops upstream are more problematic because there is no smoothing of response by a process time constant. If there are no appreciable measurement time constants, the proportional mode passes the abrupt change in the process on to the controller output. In contrast, the integral mode ramps the output, creating a gradual response that is missing in the processes.

Concept: Pipeline and inline control loops have no back mixing and hence no process time constant in the response of the outlet to changes in the inlet. What lags do exist are in the wrong place. For temperature control, thermal lags associated with heat transfer surfaces provide process time constants for changes in the cooling or heating fluid that unfortunately slow down corrections by the PID. Sensor lags and signal filters create the illusion of a process time constant. When the total loop deadtime is larger than the largest time constant in the loop, whether in the process or in the automation system, the process response is termed deadtime dominant. Such processes tend to be noisy and promote the transfer of oscillations downstream. Using less proportional action by applying a smaller controller gain and more integral action by applying a smaller reset time helps moderate the consequences of noise, interactions, nonlinearities, and short-term variability.

Details: For pipeline and inline loops, decrease the controller gain to be less than ¼ of the inverse of the open loop gain, and decrease the reset time toward a low limit of ½ of the total loop deadtime as the degree of deadtime dominance increases. See the July 2012 Control Talk blogs "The ABCs of Tuning" and "Deadtime Dominance Does Not Have to Be Deadly" for more on these factors. Equation C-13 in Appendix C of Reference 1 shows how the integral time becomes ¼ of the ultimate period as the process time constant goes to zero. If you consider the ultimate period to be twice the deadtime for this case, the integral time ends up as ½ the deadtime. Add just enough signal filtering to ensure that the fluctuations in the controller output from noise are less than the final control element deadband.

Watch-outs: Fouled electrodes, sensors that fit loosely in thermowells, and ceramic protection tubes in furnaces will add a significant measurement time constant. If the measurement time constant or signal filter becomes much larger than the deadtime, you will only have the illusion of better control because you are seeing an attenuated version of the process changes. Too large of a measurement time constant shows up on a trend chart as a decrease in amplitude and an increase in the period of the oscillations. Oscillations are more prevalent and persistent. Slow final control elements (slow variable speed drives, dampers, or control valves) can give the appearance of a time constant in the manipulated flow.

Exceptions: Temperature control via manipulation of cooling and heating in pipeline and inline equipment will have process time constants from thermal lags that may approach or exceed the process deadtime.

Insight: Inline and pipeline control loops where inlet flows are manipulated have a deadtime dominant response unless the automation system introduces a time constant in the measurement or final control element.

Rule of Thumb: Decrease the PID controller gain and reset time for loops where the deadtime is much larger than the time constant.

Tip #79: Understand How Equipment and Operating Conditions Affect Process Dynamics

Greg McMillan (Technical)

Do you lie awake at night wondering what the source of process dynamics is? Do you wonder why temperature and composition controllers tend to oscillate at low production rates and low levels? Are you perplexed about why some controllers need a gain of 0.2 and others need a gain of 20? Do you wonder why level controllers can have a controller gain of 100 without oscillating? This Tip can help prevent another sleepless night, but if that is not enough, Appendix F - "First Principle Process Gains, Deadtimes, and Time Constants" has the essential knowledge to help you fall asleep right away, especially if differential equations make you drowsy. At your next BBQ, amaze friends and relatives by explaining the effect of volumes and flows on process dynamics. Just provide recliners or hammocks as you lull them to sleep.

As a result of programming simulations to study unit operations, I learned how to set up the ordinary differential equations (ODE) for the material and energy balances. I did not have to solve them. For simulations I just needed to numerically integrate them. For real-time simulations, I could use the simplest method. For Appendix F, I just had to put the ODE in the form for self-regulating, integrating, and runaway processes. A runaway response develops when the heat from an exothermic reaction exceeds the heat removal rate. An integrating response of temperature and composition occurs when there is no discharge flow to let higher or lower temperatures or concentrations out of the vessel. Gas pressure has an integrating response when changes in pressure do not result in much of a change in vent flow. Level has an integrating response because the change in pump flow with level is negligible.

If the discharge valve is closed, the level can only go up. If there is no reaction or vaporization in a batch, composition from feed addition can only go up. If the reaction is not reversible and there are no side reactions, the product concentration in the batch from reactant addition can only go up. If an acid is added that is not consumed, the batch pH can only go down. If a base is added that is not consumed, the batch pH can only go up. Composition control loops for these one-sided responses (single direction) need to use the slope of the batch profile as the process control variable where the slope setpoint goes toward zero at the end of the batch[20].

The near-integrating process gain for slow continuous processes is equal to the true integrating process gain for batch processes, supporting the idea that slow continuous processes can be effectively treated as near-integrators.

Concept: Ordinary differential equations can be set up to solve for the process gain and process time constant. The process deadtime comes from thermal lags, mixing delays, and volumes in series.

Details: The integrating process gain for liquid level is inversely proportional to the product of the area and the density. The integrating process gain for gas pressure is proportional to the absolute temperature and is inversely proportional to volume. The integrating process gain for temperature for vessel temperature control by manipulation of jacket temperature is proportional to the product of the overall heat transfer coefficient and area and is inversely proportional to the product of the liquid mass and heat capacity. The integrating process gain for composition control of vessel concentration is proportional to feed concentration and inversely proportional to liquid mass. The turnover time for a well-mixed volume is the liquid mass divided by the summation of the mass flows from agitation and recirculation. The controller gain is inversely proportional to the product of the integrating process gain and deadtime. For well-mixed vessels the controller gain can be quite large (e.g., 10 to 50); the Lambda factor can be quite small (e.g., 0.1 to 0.02) because the turnover time is so small (e.g., 5 to 20 sec), the integrating process gain is so slow (e.g., 0.0001 to 0.00002%/sec per %), and process time constant is so large (e.g., 200 to 1000 sec). For inline temperature and composition control, the process time constant is negligible. The process gain for these plug flow volumes is inversely proportional to throughput flow. The process deadtime is the residence time (liquid mass divided by total mass throughput flow). Plug flow volumes by definition have no back mixing and no axial mixing, but may have radial mixing.

Watch-outs: For a volume to be well-mixed, there must not be any stagnant areas and the liquid volume must be completely back mixed from turbulence or agitation. The open loop response includes automation system dynamics in addition to process dynamics. The total loop deadtime is the summation of the final control element deadtime (e.g., pure deadtime from pre-stroke, deadband, resolution, and threshold sensitivity limits and equivalent deadtime from slew and ramp rates), process deadtime, transportation delay to sensor, 1½ times the analyzer cycle time, ½ of PID execution time, and the equivalent deadtime from sensor lags, transmitter damping, and signal filters. For the large process time constants in vessels and columns, 100% of all the automation system lags become equivalent deadtime. The open loop gain is the product of the final control element gain (e.g., the slope of the installed characteristic of a control valve in flow units per % signal), process gain in engineering units, and measurement gain (100% divided by the measurement span in engineering units). The open loop gain must be dimensionless.

Exceptions: The equations in Appendix F do not take into account a change in phase. The equations for mass transfer can be set up similarly to heat transfer except that the driving force is a difference in concentrations or pressures (component vapor pressure minus vessel pressure) rather than temperatures, and there is a mass transfer coefficient instead of a heat transfer coefficient. The loss of heat from a liquid through evaporation must be included in the energy balance. Crystallization can be treated as a reaction but population balances should be added to model the number of crystals in crystal size classes.

Insight: Relatively simple material and energy balances can be developed to provide first principle relationships for understanding the effects of equipment design and operating conditions on process dynamics.

Rule of Thumb: Work with the process engineers to set up the ordinary differential equations in a simplified form that enables the estimation of process dynamics.

Tip #80: Use Sliding Stem Valves for Tighter Control[11,15]

Greg McMillan (Technical)

For pH control, where the requirements for precision are extraordinary, the reagent flow rates are so low that I was never tempted to use rotary valves other than as expendable on-off valves for pulsing. I once used small solenoid actuated ball valves for pulse duration control because the frequent stroking would wear out any other type of valve and the ball valves were cheap and relatively rugged, with little propensity to plug despite slime and solids in the reagents. For throttling control, some rotary valves can come close but none can achieve the precision of a sliding stem valve that was specifically designed to be a control valve. If I had used a rotary valve in the pH control application, the limit cycle would have been much larger than the allowable error around setpoint because of the extremely high pH process gain (e.g., steep titration curve around setpoint).

A sliding stem valve with a diaphragm actuator and digital positioner should have a precision (deadband, resolution, and threshold sensitivity) of less than 0.2%. Even the best rotary valves usually have a precision that is twice as large, which is pushing the 0.5% precision limit I deem as necessary for even less important loops. Some suppliers claim that rotary valves are better for control. I guess they are thinking about flow capacity and openness of flow path rather than precision of control. Precision sets the limit cycle amplitude and the rangeability (Tip #65).

A sliding stem valve has a direct connection between the actuator shaft and the internal flow element trim (e.g., plug). The motion is linear and direct. There is no backlash from shaft connections, translation from linear to rotary motion, linkages, or rotary element seals (e.g., ball and butterfly seals). The only stiction is from valve packing and the seating of the plug in the trim. A diaphragm actuator and digital positioner have a threshold sensitivity of better than 0.1%.

Sliding stem valves offer a better installed characteristic than that of rotary valves. The installed characteristic for rotary valves tends to become too flat at upper valve positions (i.e., more open); valve gain becomes too low. Sliding stem valves also offer more choices for inherent trim characteristics.

The most precise sliding stem valve is a roller diaphragm valve, designed for low flow, viscous flow, and sanitary service. The flow is forced to be laminar.

Concept: Sliding stem valves inherently have the best precision. Nearly all processes have limit cycles from deadband, resolution, and threshold sensitivity limits. The limit cycles from sliding stem valves can be less than the process and measurement noise and the historian data compression. However, the flow path is tortuous, with pockets and crevices that can accumulate process fluid or solids. Plugging and erosion can occur. The roller diaphragm design solves these problems but is only suitable for low flows.

Details: Precision determines the limit cycle amplitude and rangeability of control valves. Use sliding stem valves when size permits and where erosion and accumulation of solids do not cause a maintenance or safety problem. Use low friction packing. Use balanced trim to reduce seating friction if leakage is not a problem. The precision of the valve with a diaphragm actuator and digital positioner should approach 0.1% for the entire throttling range. Use a roller diaphragm valve for low flows in sanitary service (e.g., food and drug applications) and for low flow rates of viscous reagents (e.g., 98% sulfuric acid). Add an on-off valve to provide tight shutoff (isolation).

Watch-outs: Piping valve manufacturers may offer a piping globe or pinch valve as a sliding stem valve. Excessively tightened stem packing, lined valves, and soft seats can cause excessive stiction. Marginally sized actuators can cause stiction and erratic action near the seat (Tip #85). Sliding stem valve sizing should include the loss of capacity if flashing occurs. Special trim may be needed to prevent cavitation. Valves inadvertently installed in the "flow to close" direction will behave erratically near the seat due to the "bath tub stopper" effect.

Exceptions: Some highly reactive monomers, such as hydrogen cyanide, will exothermically polymerize in crevices and stagnant areas, creating a dangerous safety hazard besides a maintenance problem. Sliding stem valves cannot generally be used on slurries. Applications with a large, erosive, or corrosive flow require a rotary valve.

Insight: Sliding stem valves from a control valve manufacturer offer the tightest control.

Rule of Thumb: Use a sliding stem valve with diaphragm actuator and digital positioner whenever size and process conditions permit.

"A sliding stem valve has a direct connection between the actuator shaft and the internal flow element trim."

Tip #81: Select the Best Type and Size of Actuator for Tight Control[11,15,16]

Greg McMillan (Technical)

Since early in my career, I have always been glad to see a diaphragm actuator rather than a piston actuator on a control valve, so I was surprised to read a two-part article in the 1990s by an experienced consultant who maintained that control valves should use piston actuators. The article had excellent information and shared my view that the use of positioners did not cause problems on fast loops. Why, then, did we have dramatically different opinions as to whether to use a diaphragm or piston actuator?

The author of the article was focusing on the advantages of piston actuators in terms of being smaller, faster, more powerful, and stiffer. He did not consider diaphragm actuators have a threshold sensitivity that is one-fifth to one-tenth that of a piston actuator, have no O-ring seals to wear out, and boosters for diaphragm actuators don't need a deadband. The author of the article may not have been so adamant about piston actuators being best if he had recognized that "flow to close" valves are rarely used, balanced trim reduces the actuator force needed at shutoff, on-off valves should be used instead of control valves for tight shutoff (isolation), the actuator and spring should be sized to provide at least 150% of the worst-case thrust or torque requirement, a booster can easily be added to speed up valve response, and all tests should be done with small step changes.

Until recently, step tests predominantly used a step change of 20% or larger. For large step changes, the piston actuator's poorer threshold sensitivity is not apparent. In addition, differences in speed of response are more accentuated for large steps because of the differences in slewing rate. A control valve in service is normally responding to step changes of less than 0.5% with each execution of the PID; larger step changes are only made for large setpoint changes and for very fast disturbances.

Until the 1990s, there was little testing of the ability of a valve to respond to small signal changes and to prevent limit cycles. The result was a proliferation of cheap spool type positioners (Tip #82) and on-off valves (Tip #84) that looked fine for 20% changes but were completely inadequate for 0.5% changes. Because positioners did not offer readback of actual position, the problem remained largely unknown. High historian data compression due to high data storage costs in the 1980s and 1990s helped to hide the problem.

The ISA technical report on testing the response of control valves, issued in 2000 and updated in 2006, was a huge step forward. ANSI/ISA-TR75.25.02 showed lost motion for step sizes less than 2% and an order of magnitude increase in response time when the step size became less than 0.5%. Valve manufacturers and model numbers were not identified but it is safe to assume that the worst ones were tight shutoff valves with piston actuators and spool type positioners. The valves with the best response were most likely low friction valves with diaphragm positioners and high gain relay positioners.

Today, the maximum air pressure of diaphragm actuators has increased from 35 psig to 100 psig, which provides more thrust and torque, and digital positioners now offer better sensitivity and more flexibility in tuning. External-reset feedback in the PID can be used with fast position readback to prevent a PID output from changing faster than the valve can respond to large upsets.

Concept: A diaphragm actuator offers the best threshold sensitivity. 100 psig air pressure models are now offered, overcoming an earlier disadvantage of diaphragm actuators. A double acting or spring return piston style actuator is the next best choice. Marginally sized actuators of either type will cause erratic behavior, deadband, and stick-slip.

Details: The threshold sensitivity is about 0.1% for diaphragm actuators, 0.5% for double acting piston actuators, and 1% for link arm, rack and pinion, and scotch yoke piston actuators. The link arm and scotch yoke type have excessive deadband. The slip of a rack and pinion actuator increases as the teeth wear. Piston seal wear causes the performance of piston actuators to decrease over time. Use diaphragm actuators and, if necessary, double acting or spring return piston actuators. Marginally sized actuators may cause greater limit cycling and erratic behavior near the seat, including more overshoot on breakaway. Size an actuator to provide at least 150% of maximum thrust or torque requirement.

Watch-outs: The complete valve assembly, including actuator, shaft connections, links, shaft length and diameters, packing friction, and seating and sealing friction, must be right for the valve to be able to respond to the small signal changes needed for tight control. Putting a diaphragm actuator and/or digital positioner on an on-off valve does not eliminate the stiction and backlash inherent in these valve designs (Tip #83).

Exceptions: For liquid or polymer pressure control, furnace pressure control, and other extremely fast loops, a variable speed drive (VSD) may be needed instead of a control valve. The VSD and pump must have sufficient discharge head at minimum speed for highest static head, a high resolution input card, and no speed rate limit or signal dead-band in setup. For very fast and very high process pressures, a hydraulic actuator may be needed. These actuators are a last resort because of the special maintenance requirements and the propensity for leaks. Small electronic stepper motor valves are capable of modulating extremely small changes (0.0005 inch) at fast speeds (100 millisecond time constant). These valves are limited to small sizes (e.g., < 1 inch) and stainless steel or aluminum construction[11].

Insight: Tight control requires the complete valve assembly to be designed to respond with adequate precision to changes in signal smaller than 0.5%.

Rule of Thumb: Where possible, use a diaphragm actuator sized to meet at least 150% of the maximum thrust or torque requirement.

Tip #82: Use and Tune Smart Positioners[11,15,16]

Greg McMillan (Technical)

I was encouraged to omit positioners in one of my first projects in the 1970s as lead instrument engineer. A former power system maintenance technician taking a new career path in instrument maintenance decided that positioners would add unnecessary complexity and aggravation. Fortunately, the request that valves not have positioners occurred after the project was completed. In the days of pneumatic positioners, adjusting the links and pneumatic relays required special skills. Positioners got out of calibration within six months and the lack of position readback made maintenance more difficult and the value of the positioner more questionable.

The lead instrument engineer of the system integrator for my next project proposed various ways to reduce project costs. While he had good intentions and was very experienced, all of his suggestions that were implemented created problems that had to be corrected during commissioning and after start-up. One suggestion was that we should put positioners only on selected control valves based on a frequency response study from a noted technologist that showed that boosters should be used instead of positioners on fast loops. However, I found that many rotary valves did not open until the current to pneumatic (I/P) transducer output was 30%. During startup, we put positioners on most of the valves.

When I was promoted into Engineering Technology, my areas of expertise came to include exceptionally fast loops such as compressor surge control and phosphorous furnace pressure control. These loops were so fast that they required analog or exceptionally fast digital PID execution times. On particularly large compressors, I recommended that the 24-inch butterfly surge valves have their positioners replaced with boosters (as indicated by the frequency response study mentioned above). The surge valves inexplicably slammed shut. When I went to the field the next day, the technician showed me how he could manually move a big butterfly that had a booster by just grabbing the stem, whereas the butterfly with a positioner could not be moved by hand. I realized that putting a booster on an actuator without a positioner could create positive feedback, where the booster would assist to continue movement. The booster had assisted some turbulence-induced movement to the closed position. I subsequently dug up a 1958 article (Reference 34) which showed that you should not eliminate the positioner and should put the booster on the output of the positioner with a bypass valve slightly open. A few years later, when I had specified a fast stroking time for furnace pressure control, the supplier put boosters without positioners on the big butterfly valves. In a test at the factory, I showed how I could manually move the big valves and showed them the article. The boosters were then put on the positioner output and the booster bypass valves were tuned for a fast, stable response.

Concept: Today's smart digital positioners hold their calibration and offer diagnostics and flexible tuning. The speed of response of a valve can be increased by putting a booster on the positioner output and tuning the bypass valve and positioner response settings. However, the tuning rules are not extensively documented. Field support and training schools by the supplier currently provide the required expertise to get the most out of a positioner. External-reset feedback and fast readback can prevent a process PID controller from changing faster than the valve can respond due to size or tuning.

Details: The threshold sensitivity is about 0.1% for digital positioners, 0.2% for two stage high gain relay pneumatic positioners, and 2% for spool type pneumatic positioners. Use digital positioners. Tune a positioner on the valve before installation. After installation at operating temperature and pressure, tune the positioner again. Avoid integral action. If integral action must be used, set the integral deadband in the positioner to stop limit cycles. If the valve must be speeded up, install a booster with a bypass on each positioner output that goes to the actuator. Double-acting piston actuators require two outputs and hence two boosters. Use a booster with a deadband for piston actuators. For all types of actuators, open the booster bypass valve just enough to prevent rapid cycling of the valve (e.g., 1 cycle per second). Specify a pre-stroke deadtime and slewing rate for large valves. For fast loops, test the positioner tuning to ensure that the valve response is faster than the process controller response. If the control valve response cannot be made fast enough, add an exceptionally fast position readback with external-reset feedback per Tip #85 or detune the process controller to prevent a burst of oscillations from the process controller output changing faster than the control valve can respond in the event of large setpoint changes or big disturbances.

Watch-out: Poor actuator or valve design or sizing can limit what a positioner can do (Tip #81). Positioner readback is on actuator shaft position and not internal flow element (e.g., ball, butterfly, or rotary plug) position. For on-off valves, the difference in position between the internal flow element and the actuator shaft can be as large as 8%, leading to a deadband or stick-slip of 8% from backlash and shaft windup (Tip #83). For on-off valves showing deadband and stick-slip, diagnostics and trend charts within the positioner may indicate that the deadband and stick-slip are an order of magnitude less than actual. Even the smartest and most advanced positioner will give erroneous performance results under such conditions. Thus, putting a great positioner on a control valve may not solve all response problems.

Exceptions: If a PID execution time is smaller than 0.1 seconds or an analog controller is required for good control, a variable speed drive with no deadband or rate limiting in the drive setup is needed. Loops that are too fast for digital PID and hence for control valves (analog control holdouts) are noted in the Feb. 22, 2012 Control Talk blog "What is the Best PID Execution Time?"

Insight: Readback control of valve position compensates for a whole host of unknowns in connection with valve response but a positioner, like a controller, must be tuned.

Rule of Thumb: Use tuned positioners on all control valves and add boosters with tuned bypass valves on positioner outputs as necessary to speed up valve response.

Tip #83: Don't Use On-Off Valves for Control Valves[11,15,16]

Greg McMillan (Technical)

Prior to the 1970s, you were not in much danger of getting a control valve that would not respond correctly. You simply went to your control valve suppliers and worked with them to get the best valve. In the 1970s, piping valve manufacturers decided they could sell their rotary valves as control valves by slapping on a piston actuator and a spool positioner. Because the piping valves were cheaper, already in the piping spec, and had less leakage, many users, particularly process engineers, were fooled into buying these valves. The problem escalated to the point of absurdity. Most of the oscillations in loops in the 1980s and 1990s were due to the use of piping valves designed for on-off service as control valves.

How did this error perpetuate itself? The valve specifications of the time had an entry for leakage class but no entry for the smallest change in signal that the valve must respond to. The valves were called "High Performance Valves" because they were exceptional at meeting piping specification and leakage class requirements. Who would not want to buy a "High Performance Valve" at a great price? Even the parent companies of control valve suppliers saw the opportunity and proceeded to buy the piping valve companies, so if you go to your valve supplier for a control valve now, you may be offered a suite of former piping valves.

There was no readback of actual position in pneumatic positioners, and the feedback of actuator shaft position did not show if the internal flow element (e.g., ball, butterfly, or plug) actually moved. Typical signal changes used to test a control valve were 20% or larger. For these large changes in signal, nearly all valves will respond. The ISA standard on valve response testing, ANSI/ISA-75.25.01-2000, provides a wakeup call (Tip #81), but today the message is still not clear to the average user. The absence of entries for deadband or stick-slip in a control valve specification essentially means there is no requirement that a control valve actually moves in response to a signal.

In the 1990s, I had several memorable experiences where having smart positioners fooled plant personnel into thinking there would be no control problems with "High Performance Valves." The digital positioners for ball valves in phosphorous streams, for example, all indicated that the valves were pretty good in responding to signal changes as small as 0.5%, but the process control was horrible for these loops. Testing the valves in the shop, I found that the internal ball did not move until the change was 10% or larger. The actuator shaft responded to small changes but motion was lost in the pin connections of the actuator shaft to the ball stem and in the pin connections of the ball stem to the ball. The valves were great for reliable on-off stroking and isolation, but not for control. A similar problem occurred with large high-performance butterfly valves, again selected by virtue of a plant piping spec. We had wonderful trend charts from the digital positioners saying the valves were quite responsive. Only after I put a travel gauge on a butterfly disc did we realize that the disc was not moving.

Concept: Despite digital positioner tuning and diagnostics, on-off valves are not throttling valves because of high friction in sealing surfaces for tight shutoff, pinned connections resulting in backlash, and shaft windup. The type of piston actuators and positioners often used add to the problem.

Details: Use on-off valves to meet leakage specifications for isolation. Use control valves for throttling, from companies whose heritage is process control and not piping. Specify a % deadband and % stick-slip (resolution or threshold sensitivity) for a specified throttle range. Put the tight shutoff leakage class on the on-off valve specification and not on the control valve specification. Use v-notch ball valves and toothed butterfly valves for better installed flow characteristics. Check the catalog table of flow coefficients at low and high positions to make sure a valve has a suitable inherent flow characteristic. Plot the installed characteristic for a given ratio of valve drop to system drop at operating conditions. Try to get a change in valve gain (slope of the installed characteristic) that is less than 4:1 over the operating range. Consult the checklist in Appendix C to make sure the whole valve package will function well as a control valve and will not add variability to the loop. Test the control valve in the shop, making sure the internal flow element is actually moving per the positioner readback for small signal changes. Repeat the tests when the valve is in service with a sensitive low noise flow measurement.

Watch-outs: If you need to explain valve backlash, deadband, stick-slip, or resolution to the valve supplier, the valve may have originally been a piping valve. Piping and on-off valves are often sold as control valves by control valve suppliers. A supplier may not even offer a real control valve and may not know the difference. If you go out for bids on control valves, the lowest bid will probably be valves originally designed for on-off service. Valves supplied with packaged equipment will usually be the least expensive valves and will have exceptionally poor response. "High Performance Valves" are not high performance in terms of throttling. Operators and maintenance technicians are justifiably concerned about leakage, and you need to make sure you specify on-off valves that will address these concerns. The control valves with the best throttling capability have less friction and consequently more leakage. Not all v-notch ball valves have good inherent flow characteristics. For example, one v-notch ball from what I think was a piping valve manufacturer had a flow coefficient of 0% for 0-15% signal. When asked for response test results, most control valve manufacturers will do the tests at 50% output where the friction is low and the flow characteristic is the best. Valves can get stuck due to build-up of solids, corrosion, and expansion of final flow elements at high temperatures. Periodic pulsing of the valve can keep a valve free to move.

Exceptions: For pulse width control to prevent plugging problems, an on-off valve might be better than a control valve because the on-off valve is designed for fast, large strokes.

Insight: On-off valves should not be used as control valves (for throttling) and control valves should not be used as on-off valves (for isolation).

Rule of Thumb: For throttling, use control valves supplied by a company whose heritage is process control and not piping.

Tip #84: Use Signal Characterization in the DCS to Linearize Valve Installed Flow Characteristic[25]

Greg McMillan (Technical)

Signal characterization to linearize valve installed flow characteristic has a bad reputation due to some practical implementation problems. These problems can be addressed today by doing the characterization in the distributed control system (DCS), but the stigma remains.

With pneumatic positioners, signal characterization was done by the use of a cam. Standard cams were based on linearization of the inherent characteristic, when linearization of the installed flow characteristic is really required. Custom cams had to be cut for linearizing installed characteristics. When a positioner was replaced in the middle of the night, the chances were marginal that the new positioner would have the right cam. Operators, Maintenance, and Process Technology did not like the fact that 50% output did not mean that the valve was 50% open. When a loop was in manual or when the valve needed to be stroke tested, the unknown relationship between controller output and valve position was not appreciated, to say the least. Even with the advent of digital positioners that enabled flexible and more precise adjustment of the installed characteristic, the confusion and lack of visibility in the control room remained.

The other complaint was that signal characterization increased the effect of deadband. For a change on the steep part of the installed characteristic, the change in control signal to the positioner was smaller. The change in signal could become less than the deadband. What was not realized is that for a change on the flat part of the installed characteristic, the change in signal becomes larger, reducing the effect of deadband. Also not recognized is that part of the solution was to get a better valve with less deadband. Furthermore, functionality can be added in the DCS to help deal with deadband as noted on the August 9, 2012 Control Talk blog "Deadband and Resolution Limit Cycle Causes and Fixes."

If the signal characterization is done in the DCS, all of these problems can be addressed. A standard signal characterization block in the DCS enables a process control engineer to monitor and improve valve installed characteristic. The operator interface can include features to improve the visibility of what the valve is doing. Function blocks can be added to improve the response to small signal changes that occur when the valve is operating on the steep part of the characteristic.

Adaptive tuning and gain scheduling can accomplish some of the linearization. However, the number of gain regions is limited to five, whereas a signal characterizer has 20 or more regions. Signal characterizer blocks can be cascaded to provide even greater resolution. By doing signal characterization, the adaptive tuner is freed to take care of process gain and split range nonlinearities.

Concept: The function blocks in a DCS can overcome most of the problems that originate from doing signal characterization in the field, which was particularly problematic in the days of pneumatic positioners with cams. Improvements in the control displays can provide the necessary visibility and understanding for Operations and Maintenance.

Details: Use a signal characterization block in the DCS to linearize valve flow response based on the installed flow characteristic. Control valve suppliers offer software to compute the installed flow characteristic. The signal characterizer input is the PID output (% stroke on characteristic) and the characterizer output is the signal to the control valve (% flow on characteristic). Space the XY pairs closer to provide more compensation in the areas of greatest nonlinearity. If higher resolution is needed, cascade the signal characterization blocks. Use XY plots to show the installed nonlinear characteristic, the characterized linearized characteristic, and the relationship between PID output and linearized signal. Use a flow measurement synchronized with the valve signal to provide the operating point on an XY plot of flow versus PID output. Consider updating the installed characteristic during setpoint changes and start-up. Show the valve stroke next to the PID output on the PID faceplate. Use a deadband compensator or lead-lag block on the output of the signal characterization block to help get the signal through the deadband for operating points on the steep part of the installed characteristic.

Watch-outs: Fouling of piping and process equipment can dramatically increase the frictional losses in a system, reducing the ratio of valve drop to system drop. The installed characteristic slope of rotary valves, particularly butterfly valves, can be too flat above 60% open. Valves with a linear inherent characteristic display a quick opening installed characteristic, with a slope that is first too steep and then too flat for small valve drop to system drop ratios. Note that people sometimes mistakenly use the term hysteresis for backlash. The hysteresis in a valve response is the bowing of the stroke (maximum difference in stroke) between an increasing and a decreasing signal. The bowing is less than the deadband from backlash and is not as problematic in that the valve stroke will immediately reverse direction. For a valve with backlash, the signal has to go through a deadband to reverse stroke.

Exceptions: Valves with excessive stick-slip or a quick opening characteristic should be corrected first before being characterized. An equal percentage trim characteristic provides a valve gain proportional to flow that compensates for the process gain, which is inversely proportional to flow for concentration or temperature control of inline equipment (Tip #79). If the PID output goes directly to the control valve instead of to a flow controller, linearization of the valve by signal characterization is counterproductive. The installed characteristic may be unknown or too variable for linearization.

Insight: Signal characterization in a DCS can solve problems resulting from signal characterization in the valve positioner.

Rule of Thumb: Use signal characterization in a DCS to linearize the installed flow characteristic of a valve and free up the adaptive tuner for higher level functions.

Tip #85: Use Linear Reagent Demand Control for Systems with Difficult-to-Control pH[32,35]

Greg McMillan (Technical)

Control operators would say to me, "Can you do something to slow down the pH changes?" For a setpoint on a steep titration curve, the pH would often zip right past the setpoint on its way from one tail of the curve to the other tail. Titration curve plots from the lab for strong acids and bases have nearly horizontal lines, affectionately called "tails," below 2 pH and above 12 pH. Between 2 and 12 pH is a vertical line, with slightly rounded corners at the tails. When the pH enters the vertical portion (the neutral region), the pH change accelerates as the pH approaches the neutral point (i.e., 7 pH at 25 °C). To the PID controller, the accelerating response looks like a runaway response. Stopping the pH at a setpoint in the neutral region is like trying to stop on a dime in an accelerating car exceeding the speed limit with your foot still on the gas.

The process time constant predicted from the material balance (Tip #79) is for the response of the acid and base concentrations. The entire nice process time constant from having a well-mixed vessel essentially disappears as the process goes from the titration curve X axis, which is the ratio of reagent to influent, to the Y axis, which is pH. For example, a strong acid and strong base can cause a 20-minute time constant for the concentration response to become a 0.04 minute time constant for the pH response.

Fortunately, systems identified as strong acid or strong base have a slight concentration of a weak acid, carbonic acid, by virtue of the absorption of carbon dioxide. A little bit of a weak acid or a weak base goes a long way. The slight absorption of carbon dioxide from simple exposure to air causes a moderation of the titration curve slope near 6 pH that reduces the change in process gain from seven to three orders of magnitude in the neutral region.

In linear reagent control, the process variable for the controller becomes the X axis instead of the Y axis of the titration curve. The controller sees a concentration response that is more linear and slower. The process time constant achieved by the considerable investment in having a well-mixed vessel is restored.

Concept: The use of the X axis of the titration curve can reduce the nonlinearity by orders of magnitude. The fit of the control system curve to the lab curve is best measured in terms of how well the slopes (process gains) match near the setpoint. Even if the fit is not perfect, the improvement can be dramatic.

Details: The greatest benefits of this approach are seen for setpoints on the steep portion of the titration curve. A steep slope on the titration curve causes an acceleration of the pH change and amplifies valve limit cycles. The pH acceleration and the oscillations from an imperfect valve upset the controller and the operators. To slow down this pH excursion and reduce the amplitude of oscillations, translate the process variable (PV) from pH (Y axis) to % reagent demand (X axis). Use a signal characterizer in a Distributed Control System (DCS). The input to the characterizer is pH and the output is the PID PV in % reagent demand. Add cascaded signal characterizers to provide enough resolution near the setpoint. Ask the plant lab to provide the titration curve data points in an Excel file. Make sure there are at least 20 data points in the operating range. Get the temperature of the sample and the concentration and type of reagent used. If there are not enough data points, ask the lab to repeat the titration in the operating range with a smaller reagent dose to create more points. Change the X axis scale from a ratio of reagent to sample volume to % reagent demand. Plot the curve and the slope of the curve created in the lab and created by the signal characterizer. Zoom in on the steep slope regions to compare the nonlinearity. Adjust the signal characterizer for a better match of the slopes near the setpoint. Provide the ability for the operator to enter a setpoint in pH and to read a PV in pH. Display and trend the linear reagent demand setpoint and PV. To facilitate operator understanding, add XY plots to the displays with the ability to zoom in on the control region. Train operators using a virtual plant so they become confident in the use of a linear reagent demand controller. Use solution pH temperature compensation besides the standard electrode temperature compensation of pH measurements. The reference temperature is the sample temperature when titration was done. The use of linear reagent demand control does not eliminate the need for the precise, high rangeability control valves justified and obtained per Tips #65 and #80-84. The need for linear reagent demand control is often greatest on the first stage of neutralization because this stage must cover the most ground and must deal with the majority of the disturbances. The last stage should be designed with a greater volume, better mixing, and more precise valves to keep the pH close enough to setpoint to reduce the nonlinearity seen by the controller. Use an adaptive tuner to compensate for changes in the titration curve.

Watch-outs: Temperature changes will change the water, acid, and base dissociation constants and consequently the solution pH and the titration curve slope. Chemists will sometimes use a weaker reagent to make the titration easier. For processes not exposed to air, the exposure of the sample to air and to humans exhaling carbon dioxide may cause enough carbon dioxide absorption to change the curve. The absorption of carbon dioxide and ions from the beaker and reference electrode becomes more problematic as water purity increases (e.g., boiler feed water, condensate, and deionized water).

Exceptions: For systems going from a steep part to a flat part of a titration curve or for titration curves that dramatically change in terms of slope, signal characterization is not beneficial.

Insight: Linear reagent demand control reduces pH nonlinearity and the amplification of limit cycles.

Rule of Thumb: Use linear reagent demand control when going from a flatter part to a steeper part of a titration curve for tighter control and less wear and tear on the valves and the operators.

Tip #86: Test the pH Response Time to Determine the Health of a Glass Electrode[25,32]

Greg McMillan (Technical)

There is not a whole lot of current information on the response time of pH electrodes. Tests done in the 1960s for a clean, healthy pH glass electrode showed that the time constant was smaller for a positive pH change, a higher velocity, and a higher degree of buffering. For a buffered solution at 4 fps fluid velocity, the time constants were 0.75 and 1.5 sec for a positive and negative pH change, respectively. At 0.5 fps the time constants became 3 and 12 seconds for the same electrode and solution.

A case history published in 1990 showed that the time to reach 98% of the final response (the "response time") deteriorated from 10 seconds to 7 minutes for a 1 mm slime coating. The time to reach 98% of the final response is four time constants plus the deadtime. The use of a 98% response time can be misleading and difficult because pH electrodes typically have a much slower approach to a final value above 90% than a first order classical response. Waiting to 98% means that the test time is greatly extended and the inferred time constant may be grossly overestimated. Also, a small amount of noise can lead to inconsistent results.

As far as the controller is concerned, what happens at the beginning of the response is the most important. I prefer to use an 86% response time for a faster test and a more accurate time constant that is less sensitive to noise. The 86% response time is used for control valve response testing in the ISA-75.25.01 standard for some of the same reasons.

Test results for a glass electrode prematurely aged by exposure to high temperature showed an increase in the 86% response time from 10 seconds to 50 minutes[36]. Dehydration, abrasion, and chemical attack can also cause a large increase in response time. An increase in the sensor response time (lag time) slows down the response of the loop to upsets. In the study cited, a 1 mm of slime coating increased the amplitude of oscillations by a factor of 10 and the period doubled.

Aging of a glass electrode also shows up as a decrease in efficiency; that is, span. However, this effect is not as detrimental to loop performance. The loss in span causes a decrease in process gain but there are much larger influences on the process gain, such as the shape of the titration curve. For a setpoint at 7 pH (the zero millivolt point) there is no effect on operating point. The actual pH is 7 when the measurement is at setpoint. As the pH setpoint gets further away from 7, the effect becomes larger. However, standardization by the analysis of a grab sample will compensate for the difference between the actual pH and the setpoint, reducing the effect efficiency back to a process gain. Losses in efficiency are less problematic than offsets from the reference electrode (Tip #87).

Concept: A thin coating on the glass and/or aging of the glass will show up as a huge increase in the sensor time constant. (The 86% response time is the electrode deadtime plus two time constants.) A large sensor lag causes deterioration of loop dynamics. A measurement of the response time by making a setpoint change or by putting the electrode in different buffer solutions provides a sensitive indicator of the health of a glass electrode.

Details: When buffers are used to calibrate an electrode, estimate the time to 86% of the final response. The electrode deadtime is usually negligible. Use the data historian and trend chart in the DCS to estimate the response. If you have a wireless gateway, use a wireless pH transmitter instead of a lab pH meter to get the calibration data into the historian. Errors of less than 10 seconds in the estimate are not important because glass electrode measurement problems show up as large changes in the measurement time constant. Use setpoint changes to measure the response time of a glass electrode. If you have multiple electrodes, the increase in electrode time constant will show up as a nearly constant time shift between the response curves. A decrease in electrode efficiency shows up as an increasing time shift. Remove and clean or rejuvenate, if necessary, the slowest-responding electrode(s). Use a dilute 5% hydrochloric acid (HCL) solution to remove alkaline deposits and strip away the outer, aged layer of glass to rejuvenate an electrode. Use a dilute 1% sodium hydroxide (NaOH) solution to remove acidic deposits. Use a detergent solution to remove organic deposits (oil and grease). The household rules of cleaning solutions to remove stains may be applicable to cleaning electrodes. More tenacious deposits may require a solvent. Be careful to avoid solvent attack on the sensor's O-rings and seals. Limit exposure time to prevent contamination of the reference junction and chemical attack on the glass by the cleaning solution. To minimize coating while in service, ensure that the fluid velocity past the electrode is greater than 5 fps and that the protective shroud provides exposure of the glass surface to the flow stream unless the fluid is abrasive. Use high temperature glass to prevent premature aging from exposure to temperatures above 40 °C.

Watch-outs: The equilibration of reference potentials may make finding the final response difficult, particularly as the measurement electrode response gets slower. The extremely long equilibration time of some solid-state references will make measurement of the 98% response time thoroughly inconsistent. The measurement may appear to drift or never reach the buffer solution pH despite repeated calibration adjustments for both offset and span. The % of final response is often not given in response time statements. Sometimes people mistakenly use response time when they mean time constant. Because the number of time constants is dramatically different between 63% and 98% (1 versus 4 time constants), the source should be questioned as to the % of final response.

Exceptions: For loops subjected to frequent changes in setpoint that are not corrected by an upper level loop, a loss in efficiency (span) may become more important than an increase in sensor lag. Electrodes that have an extreme loss in efficiency are in danger of becoming dead electrodes and must be replaced even if they are fast-responding.

Insight: The sensor response time is a sensitive indicator of the effects of coatings and aging on glass pH electrodes.

Rule of Thumb: Use the 86% response time to determine when to clean, rejuvenate, and/or replace the glass electrode.

Tip #87: Monitor the pH Offset to Determine the Health of a Reference Electrode[25,32]

Greg McMillan (Technical)

The reference electrode should provide a fixed reference potential and a path of electrical continuity from the internal reference element via a porous surface or aperture between the reference electrolyte and the process called a liquid junction. Coating of the reference junction can cause slow equilibration. Process ions, such as cyanide, can cause precipitation of silver ions, which will plug the reference junction. Plugging of the reference junction can isolate the internal reference electrolyte, causing a loss of electrical continuity. Contamination of the reference junction by process ions getting into the junction can cause a significant change in the liquid junction potential. If the electrolyte around the internal element is poisoned by entering process ions, the change in reference potential will be so large that the electrode will be useless. For such applications a solid-state reference or a different electrolyte must be used. All of these conditions will show up as an offset whenever a calibration check is done.

Coating and plugging of the reference junction will also show up as an increase in the reference electrode resistance. Because the reference electrode resistance is relatively small (kilohms), a reference resistance diagnostic can provide an early warning of junction coating and plugging problems. This is not the case for the glass electrode because the glass resistance is so high (Megohms). For glass electrodes, response time is a much more sensitive indicator of coating (Tip #86).

Suppliers prefer that the electrodes always be removed and buffer calibrated because the conditions and procedure are well known and are even automated in smart transmitters. The buffer is a primary calibration standard, whereas a lab pH meter is not. For applications demanding accuracies of better than a 0.1 pH, buffer calibrations may be periodically necessary.

I have always preferred an online standardization adjustment, where a grab sample is taken to make a zero adjustment to compensate for an offset. The electrodes do not have to be removed and the installation does not have to be isolated, flushed, and drained. An online standardization adjustment eliminates glass damage during handling, upsets to the reference junction equilibrium and the process, and not seeing the effect of process temperature and ionic strength on pH. However, the grab sample pH changes as the sample cools off, gets contaminated, or changes in composition. Pure water is easily contaminated by absorption of glass ions or carbon dioxide. Process fluids can change composition from reaction or vaporization or escape of dissolved gases as bubbles. Dissociation constants change with temperature. A rapidly changing process pH is not suitable for standardization because the exact time the sample was taken becomes critical. Hazardous samples require extra handling precautions and a lab pH measurement must be done under a fume hood.

Maintenance personnel often chase calibration adjustments by responding to temporary deviations. Installing an online referee electrode can eliminate unnecessary calibrations. The installation of three electrodes and middle signal selection (Tip #88) creates a reliable referee via the middle value, which offers many advantages.

Concept: Buffer calibrations provide the best calibration under standard conditions but do not take into account temperature, ionic strength, and reference junction equilibrium effects.

Details: Process streams with high ionic strengths, low water content, and changing process temperatures will have a pH measurement that is highly dependent upon operating conditions other than hydrogen ion concentration. The reference junction potential will be large and the reference electrode will take time to equilibrate with process conditions, causing a drifting pH measurement after removal and reinstallation. Some solid-state electrodes in high ionic strength streams take a day or more to equilibrate. For difficult process streams where the hydrogen ion activity or dissociation change, do an online field standardization adjustment. Take a grab sample and immediately measure the sample pH in the field if the sample vapors do not pose a hazard. If possible use a wireless pH transmitter instead of a lab meter to get the results into the DCS historian. Verify the accuracy of the test electrode by buffer calibration before using it for standardization. Use the middle value of three electrodes as a referee to determine when calibration needs to be done. Short-term fluctuations should be ignored and the effect of deviations due to differences in response time noted. For high accuracy applications and where validation procedures are required, such as with bioreactors and pharmaceuticals, do a two-point buffer calibration, making a span and zero adjustment. Do more frequent buffer calibrations as electrodes near the end of their life expectancy. Measure and record the measurement response time (Tip #86) for all buffer calibrations. Record the calibration adjustments for the electrode. Smart electrodes can store calibration history in an electrode chip. Monitor offset as indicative of a reference problem. Monitor reference electrode resistance to get an early warning of reference coating and plugging. If the reference electrode is coated, plugged, or contaminated, replace or refurbish it. If the reference electrode has a removable junction, replace the junction and electrolyte.

Watch-outs: A sample can change composition and temperature between the time the sample is taken and the time the sample is measured. The process can change composition and temperature between the time the sample is taken and when the calibration adjustment is done. A drifting reading in a buffer calibration or upon installation is an indication that the reference junction is still equilibrating.

Exceptions: A broken or cracked glass electrode or a broken wire or loose connection will cause the pH reading to be fixed near 7 pH. A standardization adjustment will not reveal that the offset is due to a damaged glass electrode. Monitor glass resistance to detect broken or cracked glass.

Insight: The health of the reference can be monitored online with less disruption to the electrodes and the process by reference electrode resistance diagnostics and standardization in the field.

Rule of Thumb: Use a smart transmitter and smart electrodes and do standardization and/or periodic buffer calibrations based on electrode life expectancy and application conditions and requirements.

Tip #88: Use Middle Signal Selection to Improve pH Measurement Reliability[25,32]

Greg McMillan (Technical)

Biopharmaceutical batches are often worth more than a million dollars. The product quality of these batches is sensitive to changes of 0.1 pH and the steam sterilization of electrodes reduces the life and accuracy of a sensor that under the best of conditions is the weakest link in a control system. Yet people looking at project costs will balk at using three electrodes because of hardware and installation costs, despite the fact that middle selection inherently protects against a single electrode failure of any type and reduces unnecessary maintenance. A million-dollar batch is put at risk by trying to save a few thousand dollars. Large continuous plants may have production streams worth several hundred thousand dollars per hour, with product quality dependent upon pH. Millions of dollars of production are routinely put at risk by the use of single electrodes for pH measurement.

The only way I can explain this is a focus on up-front costs, a lack of understanding of Tip #55, and instrumentation limitations. People may think pH electrodes are as accurate, stable, and reliable as a transmitter, when in reality electrode life can vary from hours to months, and drift can vary from 0.1 pH per month to 0.1 pH per hour.

Human nature often wants to believe the best. We see the best-case scenario (e.g., lab conditions) from advertisements and sales presentations. With just one electrode, we want to believe the reading is correct. With two electrodes, we pick a favorite and discount the other. With three, it is possible to develop realistic understanding of measurement capabilities. At Monsanto and Solutia, pH control loops had three electrodes and middle signal selection.

Middle value selection also inherently protects against the effects of a single electrode failure of any type. If an electrode goes high or low due to a reference electrode offset, one of the other two electrodes will be selected. If a glass electrode becomes sluggish, one of the other two electrodes will be selected for a pH excursion. If a glass electrode, wire, or termination is broken, one of the other two electrodes will be selected even if the broken electrode causes a pH reading that is at the setpoint (e.g., 7 pH) because when the process does change, one of the other two electrodes will be chosen.

Middle value selection will also reduce noise and spikes in the pH measurement. Unless the noise and spikes are exactly the same in two or more electrodes, the largest excursion will be ignored by selection of one of the other two electrodes. No lag is introduced into the measurement response as it is when a signal filter is added to attenuate noise.

Concept: The use of three measurements with middle signal selection will inherently ignore a single measurement failure of any type and spikes and errors in a single measurement. The middle value is the referee to detect and quantify problems such as offset and drift in the reference electrode and a loss of span or speed of the glass electrode. The knowledge gained can reduce maintenance costs. The benefit is considerable for pH measurements because the electrodes are vulnerable to offsets, coatings, failures, spikes, and noise and because batches or product streams can be worth a lot of money. The concept can be extended to any measurement where the benefits obtained provide an acceptable return on investment.

Details: Use three measurements and middle signal selection to be able to ignore problems with a single measurement. If pH is important enough to control, use middle signal selection to improve reliability, accuracy, and troubleshooting. A time delta that is fixed or is increasing between the response of an electrode and the middle value for a disturbance or setpoint change is indicative of a slow and low efficiency glass electrode, respectively. For a slow electrode, the fixed time delta is the time constant. A slow glass electrode can be caused by coating and/or aging of the glass (Tip #86). Offset and drift in the reference electrode can be caused by junction equilibration, coating, or plugging or by internal electrolyte poisoning (Tip #87). Use the understanding gained to reduce unnecessary calibration adjustments and provide predictive rather than reactive maintenance based on trends in sensor offset, efficiency, and lag time. The reduction in life cycle costs for more efficient maintenance often provides a sufficient return on investment without even getting into process benefits. Use three measurements, and use middle signal selection for any measurement that can be the probable cause of a bad batch, poor product quality, environmental violations, or a plant shutdown. The process benefits can more than pay for the additional instrumentation. Use three measurements and use middle signal selection to prevent false trips in Safety Instrumented Systems (SIS) and in critical loops that are responsible for preventing abnormal conditions getting to the point where a trip by the SIS is required.

Watch-outs: Experienced configuration and process control engineers may try to do something better than simple middle value selection. To protect against a second or third failure, take advantage of this expertise but make sure the customized technique does not interfere with middle signal selection to protect against the first failure. When a measurement failure is detected, the measurement sensor should be replaced immediately to restore the inherent protection offered by middle signal selection.

Exceptions: If process conditions are so harsh that the life expectancy of the best pH electrode is only a few days, try to find a different measurement such as conductivity or density. If an alternate means of measurement does not exist, consider using an automated retractable sensor assembly to periodically insert a single pH electrode just long enough to get a measurement. Upon insertion, wait for the electrode to stabilize before updating the pH for control. When the electrode is automatically withdrawn, the electrode should be automatically flushed and buffer calibrated. The offset and response time from the buffer calibration should be historized and analyzed to determine the health of the electrode. An enhanced PID developed for wireless (Tip #100) can be used to deal with the large time intervals between periodic insertions and subsequent updates of pH.

Insight: The use of three measurements and middle signal selection can inherently protect against any type of sensor failure and ignore noise and errors in one measurement.

Rule of Thumb: Take three measurements and use middle signal selection for pH and for any other measurement that can be a probable cause of a significant loss in process efficiency or capacity.

Tip #89: Identify Deadtime and Ramp Rate

Greg McMillan (Technical)

In Tip #70, we learned that deadtime was the key to loop performance. The Control Talk blog "The ABCs of Tuning" describes how tuning settings can be reduced to a simple function of deadtime. The *InTech* article "PID Tuning Rules" and the article's online appendices describe the fundamental relationships between deadtime, performance, and tuning.

Fortunately, deadtime is the easiest aspect of process dynamics to identify. The deadtime is the delay between a change in control output and the beginning of the resulting change in the process variable. To ensure that it is a process response and not noise or coincident unmeasured disturbances, you must ensure that the response is in the right direction and is sustained.

Once measured and confirmed, the deadtime is used by a deadtime block to create an old *PV* from a new *PV* at the block input. The ΔPV added to the current *PV* is a predicted *PV* on deadtime in the future, which opens up all sorts of opportunities (Tip #90). Convert the ΔPV to a percent of measurement scale. If you divide the $\Delta\%PV$ by the deadtime (θ_o), you have a continuous train of ramp rates ($\Delta\%PV/\Delta t$) that are updated with every execution of the block. The ramp rate can be used for a smarter integral mode and feedforward action (Tip #92). If you divide this ramp rate by the change in controller output, you have the integrating process gain (K_i) (Equation 1) that can be used for controller tuning, rapid modeling, and adapting dynamic models online (Tip #98). If you divide a steady-state open loop gain (K_o) by the integrating process gain, you have the open loop time constant (τ_o) for a self-regulating process (Equation 2). The open loop gain (K_o) can be approximated as the ratio of percent process variable ($\%PV$) to the ($\%CO$) controller output at the setpoint. If you have knowledge of another operating point ($\%PV_o$, $\%CO_o$) or know the $\%PV_o$ when the controller output is zero ($\%CO_o=0$), you can subtract these other operating point values from the values at the setpoint to create deviation variables that give a more accurate open loop gain (Equation 3).

$$K_i = \frac{\Delta\%PV/\Delta t}{\%\Delta CO} \tag{1}$$

$$\tau_o = \frac{K_o}{K_i} \tag{2}$$

$$K_o = \left|\frac{\%PV - \%PV_o}{\%CO - \%CO_o}\right| \tag{3}$$

The ramp rate for level can be used to create a rate of change of vessel level or weight for an inferential measurement of flow. The deadtime block must use a deadtime much larger than the process deadtime so that the ΔPV, and consequently the ramp rate, is much larger than noise.

Concept: The identification of deadtime and ramp rate opens up a wide spectrum of opportunities. The use of a deadtime block creates a continuous train of ramp rates as fast as the block executes.

Details: Use auto-tuner software, adaptive tuner software, or the rapid modeler composite template library block to identify the deadtime, ramp rate, and the integrating process gain or open loop gain, and open loop time constant for any large change in controller output or feedforward signal. Use a noise band to screen out insignificant changes. Make sure the identifier is looking for a change in the right direction. Use a time interval much larger than the deadtime to increase the signal-to-noise ratio when computing the ramp rate for the rate of change of level and weight for an inferential flow measurement. Also increase the time interval when computing the slope of batch processes for batch end point and cycle time optimization (Tip #96). For inverse response, the deadtime will be increased automatically and the ramp rate measured will be based on the response in the right direction. The method can be used to identify the dynamics between any process input and process output. Use the simple relationship between a true integrating or near-integrating process and a self-regulating process to convert between an integrating process gain and steady-state dynamics; that is, open loop gain and open loop time constant. A deadtime block in the identification of the ramp rate and the subsequent integrating process gain is essential to improve the signal-to-noise ratio and provide a continuous train of values. Use the identified dynamics to adapt tieback models (Tip #98). Use deviation variables to get a more accurate open loop gain. For an integrating process or runaway process models, subtract a load equal to normal controller output from the current controller output. This load stops the ramp or divergence when the current controller output balances out the load.

Watch-Outs: The deadtime and ramp rate should only be identified for setpoint changes and output changes in manual, remote output, and output tracking mode that are large enough to make noise and unmeasured disturbances negligible. The deadtime cannot be identified in automatic, cascade, or remote cascade mode if there are no setpoint changes or no injection of a known change in controller output.

Exceptions: Dynamics cannot be identified accurately for processes with a deadtime approaching the execution time of the deadtime block or the identification module.

Insight: The identification of deadtime and ramp rate can provide tuning, rapid modeling, and future values for smart reset, feedforward, and setpoint responses.

Rule of Thumb: Use a deadtime block to create a continuous train of old PV that when subtracted from the current PV and divided by the deadtime creates a continuous train of ramp rate updates as fast as the deadtime block execution time.

Tip #90: Compute and Trend a Future PV

Greg McMillan (Technical)

A change anywhere in a control loop will circle the loop in one loop deadtime. Every part of the control loop in the field and in the control room will be affected by the change within one deadtime. Loop components in the path as the change propagates will see the response before one deadtime but the effect of the change will not be seen at the point of origin until one loop deadtime has passed.

This concept is essential because people can get confused about the effect of digital measurement delays resulting from wireless update times and about controller scan times and execution times. If tests are made by making a change in a measurement right before a wireless device reports or the controller executes a scan or algorithm, there is no apparent deadtime. What you need to realize is that a process disturbance can arrive at any time within a wireless update time, controller scan time, or module execution time. Furthermore, the response of the controller has to pass through the delays in the loop to get to the point of measurement or control, and there is no guarantee that the response will arrive at the point of measurement or control right before it is ready to update. Statistically on average, the change will occur in the middle of the digital device time interval. The equivalent deadtime is ½ of the digital device time interval plus latency; that is, the time to output a result after an input is received. The latency for most digital devices today can be taken as zero. The latency for chromatographs is the full cycle time, leading to an analyzer deadtime that is 1½ times the cycle time. The phase shift and the ultimate period confirm these deadtime approximations. (For more on the importance of deadtime, review Tip #70.)

A change in controller output does not immediately change the controller's process variable (PV). The fact that nothing is seen, not even a partial response, until one deadtime in the future makes anticipation of what is going to happen difficult for humans and control systems.

A deadtime compensator can show the PV response one deadtime into the future, based on a change in controller output. A model predictive controller (MPC) can show the future PV trajectory based on the past number of MPC moves (changes in MPC output) and an experimental dynamic model of the process response to a move. What is missing in both future values is the dynamic effect of disturbances and load changes. The MPC will bias the trajectory based on the error between the predicted current PV and the actual PV, but the shape of the trajectory will not change.

A future value computed from the identified ramp rate reflects the effect of controller actions, unknowns, disturbances, and loads without any preconception of the process dynamics other than loop deadtime. For abnormal operation, start-ups, nonlinearities, batch operations, and non-stationary responses, the ramp rate method offers a better view of the future than MPC trajectories.

Concept: A future PV value can help Operations understand the effects of their setpoint and manual output changes and be more patient. A future PV value can help a PID controller optimize setpoint changes and deal with abnormal operations by various output tracking strategies (Tip # 91). A future PV value can also be used for batch cycle time and yield optimization (Tip #96).

Details: Compute a future value one deadtime into the future by multiplying the ramp rate identified with a deadtime block (Tip # 89) by a time interval. To see the effect of doing this one deadtime into the future, multiply the ramp rate by the deadtime. This computation can be simplified to the delta PV created by the deadtime block (Tip #89) added to the current value. To provide a projection further into the future to provide more anticipation, multiply the PV ramp rate by a time greater than the loop deadtime. For batch profile and end point control, the time interval must be large enough that noise in the batch profile slope calculation does not affect the batch optimization (Tip #96). For smart integral action (Tip #92), increase the time interval as necessary to prevent overshoot. For control loops with controller gains much higher than 1, the future is better seen in the future controller output (CO). The CO ramp rate is computed the same way. Scale limits and output limits set the minimum and maximum future PV and CO, respectively. Plot the future values and the ramp rates on the trend chart per the checklist in Appendix C to help the operators. Consult Tips #92, 92, and 96 to learn how to use the future PV value in process control. For stationary continuous processes at normal operating points, a future PV value from an MPC trajectory is better because the trajectory provides more information and is largely determined by previous MPC moves.

Watch-outs: Transportation deadtime and injection deadtime in a process are inversely proportional to flow rate. For pH processes with small reagent flow rates, the injection deadtime is particularly large. Batch and start-up deadtime will increase as the equipment is filled. The deadtime will also change as process conditions change during the progress of a batch or start-up.

Exceptions: The future PV value calculation will not work for deadtime dominant processes, noisy processes, and valves with excessive deadband and stick-slip.

Insight: A future PV value can make setpoint response, reset, and feedforward smarter and help operations understand the consequences of the dynamics on current actions.

Rule of Thumb: Compute, trend, and use a future PV, and, if advantageous, a future CO value.

Tip #91: Use Output Tracking to Improve Setpoint and Abnormal Condition Response[6]

Greg McMillan (Technical)

If you watch a good operator start up or deal with abnormal conditions, you see him or her put the controller output at a value that puts the process at a good operating point based on past experience. The good operator is patient and confident enough to leave the output at this value long enough for it to take full effect. We can take the best operator actions and use the output tracking function of the PID to hold these values for an optimum amount of time.

For start-up, the optimum tracking time is the rise time (the time to reach setpoint) minus one deadtime because changes are not seen until one deadtime into the future. The remaining rise time is the absolute value of the difference between the process variable (PV) and the new setpoint divided by the ramp rate (in consistent units). A future PV value can also be used to trigger the end of output tracking. For the fastest possible response, the output tracking value could be an output limit. When the future PV is just short of the setpoint, the final resting value is tracked for one deadtime[3].

For abnormal conditions, a conservative output or incremental output is tracked. This strategy is called an open loop backup because feedback action is judiciously suspended to take extreme action. The conservative output provides a step change that protects against a worst-case scenario. An incremental output is usually more efficient but causes a slower response. In practice, a combination of an initial output to get a larger, more immediate response and followed by an incremental output may be best. Output tracking is triggered when a PV, future PV, or ramp rate indicates a potentially damaging or unsafe condition or environmental violation is eminent.

To prevent compressor surge, I use an output that I know will get the compressor out of surge. Each compressor surge cycle damages seals and possibly the rotor, causing a loss in efficiency. For axial compressors, the loss in efficiency can be noticeable after just a few surge cycles. Once a compressor gets in surge, a feedback controller usually cannot get the compressor out of surge because the oscillations are normally too fast for attenuation and are very large. Making the surge control loop faster only makes resonance more likely, amplifying the surge cycles. In classic surge conditions, there is a negative swing in the flow in 0.005 seconds that is as large as the positive flow. The flow walks along the negative flow compressor characteristic curve and then jumps in 0.005 seconds to positive flow close to the surge point. This cycle repeats every 1 to 2 seconds. I detect the onset of surge by the initial precipitous drop in flow or by an operating point approaching the surge curve and use an open loop backup to position the surge valve open enough and long enough to prevent surge cycles[27].

To prevent an RCRA pH violation, an incremental output is normally fast enough and saves on reagent use. For an inline pH control system, I found that an incremental output of 0.5% per second was fast enough and saved $50K per year in reagent use compared to a 50% output for tracking[37].

Concept: Use output tracking to provide the initial change in output necessary to get a process to setpoint faster or to prevent abnormal operation, notably compressor surge or RCRA pH violations. Hold the output long enough to ensure that the full effect of the output change is achieved and the loop has stabilized. The loop may then be turned back over to feedback control to correct for unknowns and disturbances and to provide more efficient operation.

Details: If you want the PV to smoothly reach setpoint about as fast as the open loop response, track the final resting value for the rise time less one loop deadtime. The remaining rise time for tracking at any point in the setpoint response can be estimated as the absolute value of the difference between the PV and the setpoint divided by the current ramp rate. If you want the PV to get to setpoint as fast as possible, track an output limit, and, when the future PV is close to setpoint, track the final resting value for one deadtime. The final resting value can be the best value or last value of a start-up or a batch operation. For a pure integrator with no change in load, the final resting value is the output just before the setpoint change. To prevent compressor surge, use both a PV excursion past the surge controller setpoint and a precipitous drop in flow as triggers for output tracking. A controller execution time of 0.1 seconds is needed to be fast enough. The major sources of deadtime in a surge loop are the automation system components. Hold the output at a value for much longer than the deadtime (e.g., 10 sec). The deadtime in a well-designed surge loop is too small (e.g., < 1 sec) to use a future PV value to trigger output tracking. To prevent RCRA pH violations, track an incremental output when pH approaches the RCRA limit (e.g., 2 pH and 12 pH). The deadtime is slow enough (e.g., > 5 sec) even for inline pH systems (static mixers) to use a future pH value to trigger the output tracking. Use an initial tracking value that is subsequently incremented to get a more immediate recovery.

Watch-outs: Same as Tip #90.

Exceptions: For frequent and large load changes during setpoint changes, the PID controller may need to stay in automatic. In this case, consider using setpoint PID structure, a setpoint filter, and setpoint feedforward rather than output tracking to optimize the setpoint response.

Insight: The best actions of a process engineer and operator can be replicated by the controller as preemptive actions that are corrected by feedback after the full preemptive effect is achieved.

Rule of Thumb: Use output tracking to get the loop consistently started for the best setpoint response and recovery from an abnormal condition.

Tip #92: Make Reset Time and Feedforward Action Smarter

Greg McMillan (Technical)

The reset time can be approximated as a factor of the deadtime, but this dimensionless factor can range from 0.5 to 40 depending upon the type of process. The control literature has focused on self-regulating processes with process time constants in the same range as the deadtime and with little treatment of deadtime dominant processes, almost no consideration of batch processes and other integrating processes (the exception is Lambda tuning), and essentially no analysis of runaway processes. This deficiency could merit a book in itself, but the point here is that engineers and operators have confusion and uncertainty with regard to reset time settings.

I found that for deadtime dominant processes (i.e., processes with a deadtime much larger than the open loop time constant) the approach back to setpoint would be slow for load changes and would falter (i.e., flatten or hesitate) for setpoint changes. The solution was to reduce the reset time to a low limit of 0.5 times the deadtime for essentially pure deadtime processes.

For integrating processes, I realized that the reset time factor depended upon the controller gain setting. If you use the controller gain for maximum load rejection, the reset time is four times the deadtime. However, if you decrease the controller gain, you must proportionally increase the reset time, a non-intuitive effect since we are taught that decreasing controller gain always helps to reduce oscillations[1].

Reactors in highly exothermic processes (e.g., polymerization reactors) can exhibit a runaway temperature response[24,38]. The use of too much integral action can be dangerous. In some cases, a proportional-derivative mode is used. I found I had to increase the reset time to be 40 times the deadtime to prevent excessive overshoot in simulation tests.

Feedforward needs to be made smarter as well. Terry Tolliver, the world's best distillation control expert in my estimation, found that for column temperature control, flow feedforward could make the effect of a disturbance worse. A feed change could provide a temperature change in the direction to compensate for a concentration change or some other unknown disturbance that was eliminated by flow feedforward. I found in simulation tests that, in general, for temperature and concentration control, an enhanced PID could correct the flow feedforward signal to reduce the feedback correction.

Concept: A smarter reset time can be computed online, based on a future PV value and ramp rate, to provide a balance between integral and proportional/derivative mode action to suppress overshoot and reduce faltering. The ramp rate can be used to compute a rise time to get the output to come off of an output limit at the right time to prevent overshoot[6]. A future PV value and ramp rate can be used to prevent feedforward from doing more harm than good from unknowns. An enhanced PID can be set up to correct for significant errors in the feedforward signal to reduce the feedback correction.

Details: If the future PV value and ramp rate show that overshoot will occur, increase the reset time. If the future PV value and ramp rate show a lingering offset, faltering, or slowness in the approach to setpoint, decrease the reset time. Suspend the adjustment of reset time when the future PV is either too far away or within the allowable control error around setpoint. Continue the adjustment of reset time for each significant load and setpoint change. Set upper and lower reset time limits to prevent problems. Start out with tight limits until you gain confidence in the procedure. For processes where the analyzer cycle time is much larger than the process time constant, use an enhanced PID so the reset time does not depend upon analyzer cycle time. For large setpoint changes, abnormal operation, batch operations, start-ups, and surge control, compute the rise time from the ramp rate and the resulting reset time to get the PID output off an output limit so the PV ends up within an allowable error around setpoint[6]. If the future PV value and ramp rate based on both the actual response and the feedforward response are moving in the same direction, suspend feedforward action. Make a bumpless transition back to feedforward action when the future PV value shows a return to setpoint, indicating that the unknown disturbance has been accounted for by feedback action. Use an enhanced PID to improve the feedforward signal. This adaptive controller's PV is current feedback correction, setpoint is zero feedback correction, and output is a flow feedforward signal bias or gain. Determine from an analysis of manipulated flow versus feedforward flow for various conditions whether the feedforward signal has an offset or slope that needs to be corrected. Set a threshold sensitivity limit to ignore insignificant corrections. Set directional setpoint rate limits in the analog output (AO) block used to adjust the bias or gain so that correction can be made slow enough not to upset the process controller. Set low and conservative high AO setpoint rate limits. Make sure the adaptive controller has external-reset feedback set up with the AO block so that the adaptive controller output does not change faster than the rate limited AO block response. The feedforward correction may need to be faster in one direction.

Watch-outs: Lambda tuning for self-regulating processes shows the reset time as a factor of the open time constant. However, if you use the Lambda tuning settings for integrating processes and near-integrating processes, the reset time is three times the deadtime giving better recovery from disturbances. Improper controller gains can cause excessive oscillations or too slow of a response that cannot be corrected by a smart reset or smart feedforward. The controller gain for best practical load rejection can be universally approximated as an 0.5 factor multiplied by the inverse of the product of the integrating process gain and deadtime. For self-regulating processes, the second equation in Tip #89 can be used to convert to the more familiar expression of controller gain in terms of a steady-state open loop gain and open loop time constant.

Exceptions: The future PV value calculation will not work for deadtime dominant processes, noisy processes, and valves with excessive deadband and stick-slip. For processes with the loop deadtime approaching the execution time of the PID, the calculations are too late to make any correction.

Insight: Smart reset can reduce overshoot and faltering, and smart feedforward can provide a more complete and accurate feedforward action.

Rule of Thumb: Use future PV value and ramp rate and an enhanced PID to make the reset time and feedforward action smarter.

Tip #93: Use a Viewable and Adjustable Flow Ratio and a Feedforward Summer

Greg McMillan (Technical)

If you look at a process flow diagram, you quickly realize that the flows are designed to move up and down in unison. The ratios of the flows are maintained. If you double production rate, you can double the process flows as a first approximation. The utility flows for heating and boilup will probably double as well. For cooling, the flows will need to increase but the ratio may change because the heat transfer coefficient changes with cooling flow, and cooling water supply temperature may change with cooling water return temperature. As a result, you can readily recognize that the most common feedforward signal is flow and that flow feedforward control is really flow ratio control.

Controllers have feedforward multipliers and feedforward summers built-in to the PID algorithm. There is a bumpless transfer when feedforward action is turned on or off. In the literature, a traditional analysis shows that the intercept of a plot of manipulated flow versus feedforward flow is zero. The conclusion is that a feedforward multiplier should be used to correct for slope. In practice, my associate Terry Tolliver, world's best distillation control expert, and I found that a feedforward summer was a much better practical solution and that the built-in feedforward did not provide the necessary visibility and flexibility.

First of all, there are scaling problems when you use a multiplier. Second, by doing this, you introduce a nonlinear gain in the output that is proportional to flow. A cursory analysis might indicate that this is beneficial, because the process gain is inversely proportional to flow for temperature and composition control. However, for vessels and columns, the process time constant is also inversely proportional to flow. Because the controller gain is proportional to the ratio of time constant to process gain, the effect of flow on the controller gain via the process is cancelled. For plug flow volumes that are deadtime dominant, such as static mixers, the time constant is negligible and there is a slight benefit from a feedforward multiplier. The benefit is marginal because the controllers on these volumes use more integral action than gain action due to deadtime dominance (the reset time factor decreased from 4 toward 0.5; Tip #92).

Last, most of the error in a flow measurement is an offset that is best corrected by a feedforward summer and the slope (ratio) is best corrected by Operations. The summer output like the model predictive controller bias can be historized and analyzed for improvements.

The desired ratio must be accessible and the current ratio must be visible to Operations. For the start-up of unit operations, particularly for reactions and separations, flows are normally put on straight ratio control. The temperature or composition controllers are not put in automatic until the process equipment has reached normal operating conditions.

Concept: A ratio station is needed to provide Operations with the interface needed to start up and take control of equipment for abnormal operation. The interface and the historization of the actual ratio and desired ratio enable analysis and continuous improvement. A feedback correction from a process controller should provide a positive and negative bias to the ratio station output. When the bias is zero, the feedforward signal is perfect. Feedback control must take over if the feedforward signal is not usable because of flowmeter problems or a flow rate beyond the rangeability of the flowmeter.

Details: Filter the feedforward flow measurement so that noise does not cause the control valve to dither; rapidly cycle. Use a ratio station for the operator to set the desired ratio and see the actual ratio. Prevent zero or negative flows from causing bizarre ratios. Bias the ratio station output by a process feedback controller to correct for feedforward errors. Scale the feedforward bias to provide a correction that covers the desirable operating range. For example, if the operating point is 50%, a -50% to +50% bias should be provided by the feedback controller to adjust the manipulated flow from 0 to 100%. Use cascade control so the effect of valve nonlinearity is removed by a secondary flow loop for the manipulated flow. Set up the ratio and bias station so the feedback controller can be put in manual to stop feedback correction but not stop feedforward action. Allow the loop to run with only feedback action and no feedforward. If the feedforward flow is beyond the flowmeter's rangeability, go on total feedback control. If the manipulated flow is beyond the flowmeter's rangeability, go off of cascade control and onto direct throttling of the control valve. When not in cascade control, schedule the process controller gain and the ratio to account for the nonlinearity introduced by the control valve installed characteristic. Consider whether a smart feedforward (Tip #92) is needed.

Watch-out: Orifice flow measurements get noisy below 25% of the maximum flow. Vortex meter signals may drop out below 10% of the maximum flow. In the start-up of boilers, the drum level controller may have to directly manipulate the feed water valve until the feed water flow is above the flowmeter's low rangeability limit. Turning on three element level control, that is, feed water to steam flow ratio control, may need to wait till the steam flow is above the flowmeter's low rangeability limit. Make sure a flow ratio is not used as a process variable for a controller. A flow ratio is too nonlinear for PID control.

Exceptions: Poor flow measurement selection, installation, or maintenance can cause flowmeters to be too erratic or unreliable for flow ratio control.

Insight: Temperature and composition loops should be able to switch as necessary between feedforward only, feedforward plus feedback, feedback only, cascade, and single loop control.

Rule of Thumb: Set up a ratio station, feedforward summer, and cascade control system for maximum flexibility and visibility to address start-up and abnormal conditions and measurement limitations.

Tip #94: Smooth the Transition between Batch and Continuous Operations

Greg McMillan (Technical)

Processes often include a mixture of batch and continuous operations. For liquid products made in a batch operation, most of the time the batch flow is off because the drain valve is closed during the batch. When done, the batch is drained quickly. This discontinuous batch flow cannot directly feed continuous operations downstream. For traditional batch operations, where charges are sequenced and completed as fast as possible, continuous operations upstream cannot directly feed batch operations. In either instance, a feed or product tank is needed that acts as a surge volume. A level controller is often used to smooth the changes in continuous flow required to balance the batch flow changes and thereby minimize the upset to the continuous operations.

I worked in a large chemical intermediates plant that had a mixture of batch and continuous operations. The plant was running at over twice the original design capacity. There were multiple trains that could help moderate the effects of the discontinuous batch flows but due to fouling the units had to be periodically shut down and cleaned, and the surge tank volumes were marginal in the original design. Level controllers oscillated so operators tried to control the volumes manually, but this led to some extreme changes in feed rates severely upsetting unit operations.

A solution is an adapted rate limited feedforward for level control. The total batch volume (V_B) divided by the current batch cycle (T_B) time is the average batch volumetric flow (F_B) (Equation 1). The absolute value of the difference between the current continuous flow (F_c) and the average batch flow is the flow unbalance (ΔF) (Equation 2). The difference between the current surge tank level and the alarm level (ΔL) multiplied by the dimension factor (K) and divided by the flow unbalance is the time to the level alarm (ΔT) (Equation 3). The factor converts from percent level to volume. The rate limit ($\Delta F/\Delta T$) is the flow unbalance divided by the time to the alarm (Equation 4). For an unbalance that is increasing and decreasing level, the time (ΔT) is the time to reach the high and low alarm, respectively.

$$F_B = \frac{V_B}{T_B} \tag{1}$$

$$\Delta F = |F_B - F_C| \quad \text{(Batch operation upstream of a continuous operation)} \tag{2}$$

$$\Delta T = \frac{K * \Delta L}{\Delta F} \tag{3}$$

$$\frac{\Delta F}{\Delta T} = \frac{\Delta F^2}{K * \Delta L} \tag{4}$$

Concept: A feedforward of instantaneous batch flow can be velocity limited to smooth out feedforward control so that the unbalance is eliminated over the available surge inventory. The rate limit is applied in the direction of causing a level alarm. As the available inventory decreases, the rate limit increases, allowing a faster correction of the input to the feedforward summer in the level controller.

Details: Use an adapted velocity limited feedforward per the equations above to smooth out the flow manipulated by a surge tank level controller. Use consistent units for batch and continuous flows, volumes, and times. Manipulate the continuous flow by a combination of feedback and feedforward level control. For batch operations upstream of continuous operations, the feedforward signal is the instantaneous batch flow into the surge tank and the flow unbalance is the average batch flow minus the continuous flow. For an instantaneous batch flow greater and lower than the continuous flow, base the rate limit on the high and low level alarm, respectively. For batch operations downstream of continuous operations, the feedforward signal is the instantaneous batch flow out of the surge tank and the flow unbalance is the continuous flow minus the average batch flow. For an instantaneous batch flow lower and higher than the continuous flow, base the rate limit on the high and low level alarm, respectively. Other solutions include a controller gain scheduled to increase as the level gets further from setpoint, integral deadband to suspend integral action when the level is reasonably close to setpoint, and error-squared level control. If the desired continuous flow does match the average batch flow because of greater or less continuous capacity, make a judgment (Tip #96) as to whether to sacrifice batch efficiency or batch capacity for balance.

Watch-out: For multiple trains of batch operations using the same surge tank, the average batch flow must be adjusted accordingly. If the batches are staggered, the stagger time must be added to the batch cycle time. The average batch flow is the summation of each batch volume divided by the batch cycle time plus stagger time. The product of the level controller gain and reset time must be greater than the inverse of the (Tip #89) integrating process gain ($\Delta\%PV$ /sec/ $\Delta\%CO$) to prevent slow rolling oscillations (as shown in Equation A-2 in Appendix A of Reference 11). The level controller gain must be high enough when the level is high and increasing or low enough and decreasing to prevent a high and low level alarm, respectively. Using a proportional-only structure is not advisable because integral action is needed to slowly return the average level to a setpoint that maximizes available surge volume. Derivative action should not be used.

Exceptions: If a shutdown in the continuous area stops the continuous flow, there is no level control and the batch operations may have to go on hold until the continuous area starts up.

Insight: The unbalance between continuous operation flow and instantaneous batch flow can be spread over the available inventory in a surge tank by an adapted rate limited feedforward.

Rule of Thumb: Use an adapted rate limit to slow down the signal to the feedforward summer in a PID level controller without causing a level alarm in the surge tank.

Tip #95: Use External-Reset Feedback for Cascade Control and Slow Final Control Elements[4,5]

Greg McMillan (Technical)

The value of the external-reset feedback feature in PID controllers is becoming increasingly apparent to me. I got a preview of its importance when I found that 1980s vintage DCS required a fix for override control that was addressed in the next generation of DCS by this feature. This wakeup call was followed by the realization that this feature offered improved analyzer, blend, cascade, decoupling, feedforward, surge, valve position, and wireless control with a side benefit of simplified deadtime compensation.

Override controllers in a 1980s vintage DCS could walk off to an output limit when not selected despite the use of proper mode and configuration. The fix was to insert a filter in the path of each integral track signal where the filter time was equal to the reset time of the respective overridden PID. The external-reset feedback feature of the next generation of DCS provided this filter inherently in the external-reset signal by virtue of the positive feedback implementation of the integral mode. Besides solving the override problem, this key feature created many opportunities and solved the problems from a controller output changing too fast.

If the process controller output changes faster than the manipulated variable can respond, integral action will keep pushing the primary controller output beyond what is needed. When the effect of the overcompensation is seen in the primary loop, the integral action repeats the overcompensation in the opposite direction. The result is cycling.

The manipulated variable can be the setpoint of a secondary loop (e.g., a flow loop) or a signal to a final control element (e.g., a control valve, damper, or variable speed drive). The slowness of the manipulated variable response appears mostly as a rate limit due to a setpoint rate limit in the secondary loop or analog output block, integral action, or slewing rate limit. Consequently, for small errors the rate of change of the process controller output is rather slow and within the rate limit of the manipulated variable. Tuning tests and operation with small setpoint changes or small output changes show everything is fine. For large upsets and major setpoint changes, the loop bursts into oscillations that in most cases eventually decay. This occasional instability is insidious. Creative explanations are offered.

Fortunately, the solution is simple: just turn on the external-reset feedback (e.g., dynamic reset limit). The process variable (PV) of the manipulated variable must be set up as the back-calculate output signal and connected as the back-calculate input for the external-reset feedback. No retuning is required. The user is free to use directional setpoint rate limits to reduce interactions and achieve equipment and environmental protection, coordination, and optimization objectives (Tip #72) and more effective valve position control strategies (Tip #97).

Concept: External-reset feedback prevents the burst of oscillations caused by a primary controller outrunning a secondary loop or final control element. This key feature in modern PID controllers eliminates the need to retune primary controllers because directional setpoint rate limits are added to analog output blocks and secondary controllers to achieve process objectives.

Details: Turn on external-reset feedback in any controller whenever a manipulated variable may not respond as fast as the controller output. Use external-reset feedback for cascade control, setpoint rate limits in analog output blocks, and slow final control elements. Setup the PV of the secondary loop, analog output, and final control element as the back-calculate input for the external-reset feedback. Makes sure the PV value used for the back-calculate input is responding as fast as the actual PV is changing.

Watch-outs: The second, third, and fourth HART variable for position readback and speed feedback are too slow. A variable with an update rate at least twice as fast as the PID execution rate is needed. For split ranged loops, the back-calculate signals must be correctly connected to reflect which secondary flow loop, control valve, or damper is actually being manipulated.

Exceptions: If you do not have the positive feedback implementation of the integral mode, the external-reset feedback is not an option.

Insight: The positive feedback implementation of the integral mode enables the use of an external reset signal that automatically accounts for slowness in anything being driven by the PID output.

Rule of Thumb: Use external-reset feedback to prevent oscillations from a slow secondary loop or final control element and to open up opportunities to meet process objectives by setting setpoint rate limits.

Tip #96: Optimize Batch End Points and Cycle Times[20,34,40]

Greg McMillan (Technical)

When I was leaving for college, my Dad said, "Make sure you use a good grain analyzer to optimize alcohol batch time and yield." These words of wisdom would be useful in the years to come at "Purple Passion" parties with tubs of grain alcohol and grape juice, but their greatest value was seen last year (2011) in terms of ethanol plant optimization opportunities.

For that company's plants, I developed a control strategy that uses an at-line analysis of corn fermentability coupled with corn feed rate to provide an inferential measurement of production rate as the process variable for a simple flow controller. The operator could set the production rate for the front end of a plant that included a parallel train of batch fermenters. When the corn analyzer indicated that the corn fermentability (predicted yield) had increased, the ethanol production rate controller would cut back the corn feeder speed, immediately translating the increase in yield into a decrease in corn feed rate. Because corn is more than 50% of the cost of producing ethanol, the reduction in cost of goods sold (COGS) was significant. Furthermore, when an off-line analyzer indicated that a fermenter had reached an endpoint earlier or later than expected, a fraction of the equivalent change in fermentability was taken as a bias correction to the analyzer signal.

Most recently, I realized that I could use the slope of the batch profile for ethanol (ETOH) concentration to decide whether to end or extend the batch based on the values of additional yield and capacity.

For fed-batch reactors with concentration measurements, the slope of the batch profile can be controlled. The setpoint for the slope is set to match the best achievable batch profile. For bioreactors, the cell concentration profile can be used to control the glucose to glutamine ratio. A Nova BioProfile FLEX analyzer and auto sample can be used as an at-line analyzer for this optimization.

Concept: The slope of a key concentration during the batch can be used to control the batch profile. The slope of product concentration near the end of a batch and the analysis time interval can be used to calculate the value of additional capacity made available by terminating the batch and the value of additional yield obtained by waiting till the next analysis. A feed analyzer that predicts yield can provide a production rate controller that immediately optimizes yield for fed-batch processes or for the continuous front end of traditional batch processes.

Details: Use an off-line or at-line feed analyzer to compute the yield of a key raw material. For a continuous front end or fed-batch flow controller, compute a process variable that uses the predicted yield to provide an equivalent product flow rate. Use the production rate controller to immediately cut back on raw material feed rate for a measured increase in predicted yield. Use an enhanced PID (Tip #100) to deal with the update time from at-line and off-line analyzers. If the batch takes less or more time than normal to reach the end point, use a portion of the inferred change in yield to correct the feed analyzer. Consider the slope of a key concentration measured at-line or off-line during the batch as a process variable for a controller to adjust an operating condition that affects the conversion rate (or cell growth rate for bioreactors). If a periodic analysis is not available during the batch, see if the cooling rate can provide an inferential measurement of conversion rate in exothermic process chemical reactors and in fermenters for alcohol production. See if an oxygen uptake rate profile can provide an inferential measurement of cell growth rate in bioreactors for pharmaceutical production. For continuous analysis and inferential measurements, use a deadtime block to create the ramp rate (Tip #89). The deadtime for the block must be large enough to provide a good signal-to-noise ratio. From the batch profile slope (e.g., conversion rate, cell growth rate, product formation rate) near the end of the batch, compute the additional product produced in the deadtime interval or the analysis time interval for sampled measurements. Use the slope near the end of the batch to make an economic decision as to whether the batch should be terminated for extra capacity or extended for extra yield. Convert the slope to product mass flow and multiply by the analysis time interval to get the additional product mass till the next analyzer update. Divide the current product mass in the batch by the mass of each key raw material added to the batch to get the yield in terms of product per each key raw material. Divide the additional product mass per batch by this yield and multiply by the cost per unit mass of each key raw material. Sum the results to arrive at a dollar value of the additional product yielded by extending the batch. Take the current product mass in the batch and divide by the current batch time in hours to estimate the current batch production rate. Alternately, use the production rate from the flow controller based on predicted feed yield. Multiply the production rate by the profit per unit mass and the analysis time interval to estimate the value of additional capacity made available by terminating the batch. Shorten the analysis time interval near the end of the batch to make the optimization more accurate. If the analysis of a key composition or product is not available until after the batch has been transferred, the results can be cautiously used for the next batch.

Watch-outs: Bad or missing analyzer updates must be screened and not used for computing the batch profile slope. Inlet and outlet temperatures must be synchronized for cooling rate calculations.

Exceptions: Optimization using batch slope cannot be used if batch analyzer updates are too infrequent and an inferential measurement is not available.

Insight: Feed analyzers and batch profile slope can be used to optimize batch efficiency and capacity.

Rule of Thumb: Use a feed analysis of predicted yield to create a production rate controller and use a batch profile slope for optimization of batch end point and cycle time.

Tip #97: Use Valve Position Control to Optimize Process Efficiency and Capacity[5,6,7,8,33]

Greg McMillan (Technical)

There are many opportunities for process optimization by valve position control (VPC). Table 1 summarizes common examples of the use of VPC for increasing process efficiency and capacity. Efficiency is increased by reducing energy and raw material costs for a given production rate. Capacity is increased by increasing feed rates for a given efficiency. Table 2 is a summary of key PID features that make the implementation of VPC easier and more effective at achieving process objectives and dealing with process disturbances.

Optimization	VPC PID PV	VPC PID SP	VPC PID Out
Minimize Prime Mover Energy	Reactor Feed Flow PID Out	Max Throttle Position	Compressor or Pump Pressure SP
Minimize Boiler Fuel Cost	Steam Flow PID Out	Max Throttle Position	Boiler Pressure SP
Minimize Boiler Fuel Cost	Equipment Temperature PID Out	Max Throttle Position	Boiler Pressure SP
Minimize Chiller or CTW Energy	Equipment Temperature PID Out	Max Throttle Position	Chiller or CTW Temperature SP
Minimize Purchased Reagent or Fuel Cost	Purchased Reagent or Fuel Flow PID Out	Min Throttle Position	Waste Reagent Or Fuel Flow SP
Minimize Total Reagent Use	Final Neutralization Stage pH PID Out	Min Throttle Position	First Neutralization Stage pH PID SP
Maximize Reactor Production Rate	Reactor or Condenser Temperature PID Out	Max Throttle Position	Feed Flow or Reaction Temperature SP
Maximize Reactor Production Rate	Reactor Vent Pressure PID Out	Max Throttle Position	Feed Flow or Reaction Temperature SP
Maximize Column Production Rate	Reboiler or Condenser Flow PID Out	Max Throttle Position	Feed Flow or Column Pressure SP
Maximize Ratio or Feedforward Accuracy	Process Feedback Correction PID Out	50% (Zero Correction)	Flow Ratio or Feedforward Gain

Table 1 - Examples of Optimization by Valve Position Control

Concept: Valve position control (VPC) can be quickly implemented for small process optimization opportunities. An integral-only structure has traditionally been used for VPC to eliminate interaction. However, an integral-only controller is difficult to tune, can get into limit cycles, and can be too slow to prevent running out of valve; that is, a valve is on the flat part of its installed characteristic and is close to an output limit. The key PID features shown in Table 2 enable the use of PI control.

Feature	Function	Advantage 1	Advantage 2
Direction Velocity Limits	Limit VPC Action Speed Based on Direction	Prevent Running Out of Valve	Minimize Disruption to Process
Dynamic Reset Limit	Limit VPC Action Speed to Process Response	Direction Velocity Limits	Prevent Burst of Oscillations
Adaptive Tuning	Automatically Identify and Schedule Tuning	Eliminate Manual Tuning	Compensation of Nonlinearity
Feedforward	Preemptively Set VPC Out for Upset	Prevent Running Out of Valve	Minimize Disruption
Enhanced PID	Suspend Integral Action until PV Update	Eliminate Limit Cycles from Stiction & Backlash	Minimize Oscillations from Interaction & Delay

Table 2 - Key PID Features for Valve Position Control

Details: Use valve position control on valves that are the constraints to improving the capacity or efficiency of a unit operation. For loops that manipulate utility flows, use a maximum throttle position for the VPC setpoint. For expensive fuels and reagents, use a minimum throttle position for the VPC setpoint. To increase efficiency, use the VPC output to maximize a less expensive or waste fuel or reagent, minimize a prime mover (e.g., compressor or pump) discharge pressure, or maximize a chiller or cooling tower outlet temperature. To increase capacity, use the VPC output to maximize a feed rate to a fed-batch or continuous operation. Use an enhanced PID with a threshold limit to ignore inconsequential changes in valve position and to stop limit cycles. Use directional setpoint rate limits with external-reset feedback (Tip #72) to reduce interactions and provide a slow approach to an optimum and a fast getaway; movement away from the optimum for disturbances; to prevent running out of valve (valve close to output limit or on flat part of installed characteristic). Use a feedforward summer in the VPC to provide preemptive action. Use adaptive tuning to help the VPC deal with nonlinearities and changing objectives. Review the application with the process engineers. Train Operations in the use of the VPC. Work in the control room with the operators to improve tuning and make sure the VPC does not cause any operational problems.

Watch-outs: Just one abnormal condition, one bad batch, or one increase in off-spec will likely prevent the VPC from ever being used again. The VPC should start with very narrow output limits to restrict what the VPC can do. Only after the VPC is tuned and confidence is gained should the output limits be widened. The VPC will have to be retuned to deal with nonlinearities as the limits are widened. If you do not have external-reset feedback (e.g., dynamic reset limit), you may have to use an integral-only structure for the VPC to prevent interaction.

Exceptions: If unmeasured disturbances are too fast and disruptive, VPC is not feasible. If the optimum setpoint is a function of operating conditions rather than simply valve position, model predictive control is a better optimization solution.

Insight: Valve position control can be implemented in a matter of hours to start optimizing the efficiency or capacity of a unit operation.

Rule of Thumb: Use valve position control with key PID features to maximize production rate and to minimize energy use and raw material costs for a unit operation.

Tip #98: Achieve the Required Simulation Fidelity

Greg McMillan (Technical)

The best way to classify process simulators is to look at the original intent of the simulator software when it was first released, who the software developers were, who configures the models, and how the models are used.

Simulators originally designed for system acceptance testing (SAT), operator training systems (OTS), and process control improvement (PCI) focus on automation system details, such as the inputs, outputs, and dynamic relationships, that affect control system and operator performance. These OTS, SAT, and PCI simulations can be as simple as tieback models that take analog output (AO) from a control system, pass the signal through an open loop gain, deadtime, and time constant, and connect the result as the analog input (AI) signal that becomes the process variable (PV) used by the control system and seen by the operator. These models can be readily adapted to become step response models (experimental models) by the identification of the dynamics by an adaptive tuner and rapid modeler. The biases in the last equation in Tip #89 can be identified by the PV and AO at two operating points to provide a more accurate open loop gain based on deviation variables.

Increasingly, first principle models of process and equipment relationships, including reaction kinetics and mass and heat transfer rates, are being offered to improve the simulation of process dynamics. A graphical studio similar to what is employed for control system configuration is used to develop basic and advanced process modeling objects. The simulation software is developed by control system engineers and is primarily intended to be configured by automation engineers and technicians.

Simulators originally designed to create a process flow diagram (PFD) focus on process details, such as physical properties and equilibrium relationships, that affect process design. These simulators start out as simulators of continuous process operation at a steady state. The simulation software is created by process research and development engineers and is primarily intended to be configured by process engineers.

The PFD simulator is often claimed to be a high fidelity model. While this is true from a physical property and equilibrium relationship viewpoint, it is not always true from a dynamics viewpoint. In particular, the PFD simulator is often missing deadtime due to a lack of pure and equivalent deadtime from process transportation and mixing and from automation system deadband, update time, filters, resolution, threshold sensitivity, sensors lags, and wireless systems.

Concept: The prior job history of the developers and the design intent of the simulation software determine the ability of the software to meet different simulation objectives. The best tool for an automation engineer is a simulation that will accurately show the dynamic response for all scenarios, including start-up.

Details: Instead of the usual vague and subjective fidelity ratings of low, medium, and high, use the following fidelity ratings based on the functionality needed to meet operator and automation performance objectives. Physical fidelity achieved by a virtual plant (Tip # 99) is assumed.

- Fidelity Level 1: measurements can match setpoints and respond in the proper direction to loop outputs; for operator training.
- Fidelity Level 2: measurements can match setpoints and respond in the proper direction when control and block valves open and close and prime movers (e.g., pumps, fans, and compressors) start and stop; for operator training.
- Fidelity Level 3: loop dynamics (e.g., process gain, time constant, and deadtime) are sufficiently accurate to tune loops, prototype process control improvements, and see process interactions; for basic process control demonstrations.
- Fidelity Level 4: measurement dynamics (e.g., response to valves, prime movers, and disturbances) are sufficiently accurate to track down and analyze process variability and quantitatively assess control system capability and improvement opportunities; for rating control system capability, and conducting control system research and development.
- Fidelity Level 5: process relationships and metrics (e.g., yield, raw material costs, energy costs, product quality, production rate, production revenue) and process optimums are sufficiently accurately modeled for the design and implementation of advanced control, such as model predictive control (MPC) and real time optimization (RTO), and in some cases virtual experimentation.

Default tieback models automatically generated to match the configuration can achieve Level 1. Basic modeling objects, which consist of tiebacks and ramps with path logic, can achieve Level 2. Adapted tieback models with dynamics identified by the auto-tuner or a rapid modeler module (step response models) can achieve Level 3. Advanced modeling objects with first principle process models and dynamics of the automation system measurements and control valves can achieve Level 4. Adapted advanced modeling objects with parameters identified from design of experiments (DOE) or adapted by a model predictive controller (MPC) or a rapid modeler module can achieve Level 5[30,39].

Watch-outs: First principle models have a huge number of parameters. These must be visible and adjustable for improving model fidelity. The sensor deadtime may have to be increased to account for the additional deadtime observed in the process response.

Exceptions: First principle dynamic models developed for automation system design presently assume perfect mixing and identical bubble, cell, crystal, and particle size. In the future, subdivided volumes and population balances will be used to address these deficiencies.

Insight: Step response and first principle models can be adapted online to achieve much higher fidelity.

Rule of Thumb: Use tieback models, various levels of modeling objects, and online adaptation of dynamics and parameters to achieve the fidelity needed for your simulation application.

Tip #99: Use a Virtual Plant for All Dynamic Models

Greg McMillan (Technical)

Simulators for system acceptance testing (SAT), operator training systems (OTS), and process control improvement (PCI) should be part of a virtual plant, where a virtual version of the actual control system configuration and graphics, including historian and advanced control tools, interfaced to the simulator are running in a personal computer. The use of a virtualized rather than an emulated control system is necessary to allow the operators, process engineers, and automation engineers and technicians to use the same graphics, trends, and configuration as the actual installation. This fidelity to the actual installation is essential. The emulation of a PID block is problematic because of the numerous and powerful proprietary features, such as anti-reset windup and external-reset feedback.

The simulation used to develop, test, and quantify the benefits of process control improvements can be used as an OTS to help increase onstream time by operator buy-in and understanding particularly for advanced process control (APC). The supportive participation of specialists in the control room is critical as the inevitable questions arise as to what the APC system is doing. An APC system may appear to be taking actions that seem wrong to the operators due to human limitations in understanding complex relationships, delayed effects, and trajectories.

The virtual plant's processes must be able to be speeded up so that simulations for slow processes can be run within a reasonable time frame (e.g., 1 hour). The speedup can be obtained by increasing kinetics, mass transfer rates, and heat transfer rates and increasing the step size used in the integration of the differential equations for the mass and energy balances. For bioreactors, the speedup requirement is extraordinary because batch cycle times are 7 to 10 days. For bioreactors, the mass transfer rates, heat transfer rates, cell growth rate, and product formation rates are speeded up by a factor of 10. Material and energy balances are speeded up by increasing the integration step size by a factor of 50. The effect is multiplicative, so the total speedup is 500 times real time and a simulated 10 day batch is completed in less than an hour. The flow ranges for gases, nutrients, and reagents must be increased by the same factor as the transfer rate and kinetic speedup. The integrating process gains are increased and the process time constants are decreased by the same factor as the material and energy balance speedup factor. If the loop deadtime is not affected by speedup, the controller gains must be decreased by this same factor to account for the change in integrating process gains or time constants.

Concept: The actual displays and trend charts files used in the control room must be copied and the actual configuration downloaded into the virtual plant to provide physical fidelity. The models are speeded up faster than real time to ensure scenarios don't take too long. The corresponding flow measurement spans and final control element capacities are increased by the same factor as the kinetics and mass transfer rates and heat transfer rates factor. The controller gains are decreased by the same factor as the integration step size factor if the deadtimes are not affected by speedup. The controller gain remains unchanged for step response models where both the deadtime and time constant are speeded up. The controller reset time that is proportional to deadtime is speeded up.

Details: Use step response and first principle models to achieve the fidelity needed (Tip #98) in a virtual plant. Use the "Checklist for a Virtual Plant" in Appendix C. Use speedup factors as necessary so entire simulations take less than an hour. Increase the flow measurement span and final control element capacity by the same factor used to speed up kinetics, mass transfer rate, and heat transfer rate in first principle models. Increase the integration step sizes in these same models to speed up the process time constants and integrating process gains. Decrease the controller gains by the integration speedup factor, assuming that deadtime does not change with speedup. The controller gains are not affected by kinetics and transfer rate speedup if the flow measurement and final control elements are scaled up accordingly. The total model speedup is the kinetics and transfer rates speedup factor multiplied by the integration speedup factor. For step response models, decrease both the deadtime and the time constant so the controller gain does not change with speedup. For these step response models, decrease the reset time with the speedup factor since the reset time is proportional to the deadtime. See the July 12, 2012 Control Talk blog "The ABCs of Tuning" to better understand the implications of speedup on tuning.

Watch-outs: The proper simulation of deadtime is notoriously difficult. If the step response deadtime is speeded up, module execution times and measurement sensor responses must also be speeded up unless these are factored into the speedup of the deadtime.

Exceptions: The real-time simulation of momentum balances and water hammer may not be possible because these require exceptionally small integration step sizes. It may be possible to simulate surge by slowing down the dynamics by scaling down the integration step size and implementing the model in a module with an execution time of 0.1 sec.

Insight: A virtual plant provides a level of physical fidelity by the use of the actual plant's control system and operator interface that is crucial for all types of training and all levels of process control improvement.

Rule of Thumb: Use the actual configuration, displays, and trends that will be in the control room in a virtual plant for the training of operators, maintenance engineers, process engineers, and control engineers and the development, testing, and prototyping of process control improvements.

Tip #100: Use an Enhanced PID for At-Line and Off-Line Analyzers

Greg McMillan (Technical)

I have always favored online analyzers because they offer a continuous measurement with minimal deadtime. These analyzers usually have sensors in a pipeline or vessel to directly measure a property of the process, such as capacitance, conductivity, color, density, oxidation reduction potential, particle size, pH, turbidity, and viscosity. Expensive analyzers such as mass spectrometers that serve many vessels do not offer a continuous measurement because a sample or slip stream is cycled through them.

Despite the promise of new technologies, such as Near Infrared (NIR), the workhorse of the process industry is still the gas chromatograph, as discussed in the Dec. 2011 Control Talk column "Analyze This!" The deadtime introduced by these at-line analyzers is 1½ times the analyzer cycle time (Tip #90) plus the sample transportation delay and multiplex time. The total deadtime from a chromatograph can range anywhere from 30 to 60 minutes.

At-line analyzers may be more accurate but present more challenging maintenance and control than online analyzers. At-line analyzers require sample system maintenance and special technician expertise for troubleshooting. Failure to update causes the controller to ramp off toward an output limit when an analyzer measurement is used for closed loop control.

Most raw material and batch compositions are measured off-line in the laboratory. The time it takes the operator to take the sample and the lab technicians to do the analysis and enter the result is typically long and variable. As a consequence these results are essentially useless for closed loop control with a traditional PID.

We have seen the consequences of deadtime (Tip #70 and #71). A traditional PID requires the controller gain to be decreased and the reset time to be increased as the deadtime is increased. Composition controllers that use at-line analyzers must be tuned much slower than temperature controllers on the same equipment. Composition controllers are usually relegated to trimming temperature setpoints. The temperature controller takes care of most disturbances and the composition controller slowly corrects for temperature sensor drift and changes in the equilibrium relationships between temperature and composition. Most of the applications are on slow unit operations, such as distillation that have a time to steady state of 10 hours or more. An enhanced PID opens the door for the use of at-line analyzers and even off-line analyzers on even fast processes and improves the loop performance for all analyzers.

Concept: An enhanced PID does not compute the integral mode contribution until there is an update of the measurement or there is a setpoint change. The exponential response of the external-reset signal is used. The contributions from the proportional and derivative modes also do not change. Consequently, between analyzer updates or for a failure of an analyzer to update for a constant setpoint, there is no change in the controller output. When the deadtime from the analyzer system is greater than 95% of process response time, the controller gain can be as large as the inverse of the open loop gain, and the reset time can be based on the original process dynamics. For the accurate identification of the open loop gain, the controller can make a single correction that compensates for a setpoint change or an analyzer update. Long and variable update times from off-line analyzers do not affect the tuning.

Details: For analyzer applications, use an enhanced PID developed for wireless that utilizes external-reset feedback. If nearly the full process response is seen within the analyzer update interval (analyzer update time > 95% of process response time), the controller gain can be increased toward a maximum that is simply the inverse of the open loop gain. The open loop gain is the % change in measurement divided by the % change in controller output after all transients have died out (Tip #89). For cascade control, the open loop gain takes into account the effect of secondary controller scale, process gain, and the analyzer primary controller scale. The reset time can be based on the process deadtime. Use feedforward control for measured disturbances since there is little to no attenuation by feedback control because the effect of the disturbance is not seen until after the analyzer deadtime. Feedforward signal changes immediately change the output. If the analyzer fails to update, the enhanced PID output will not change.

Watch-outs: A bizarre analyzer value or an upscale or downscale failure must be screened out. While the enhanced PID simplifies the tuning, the response for unmeasured disturbances still depends upon the total loop deadtime. For fast unmeasured disturbances, the peak error is the open loop error, that is, the error that would appear if there was no feedback control. The minimum integrated absolute error is the peak error multiplied by ½ of the total loop deadtime. For setpoint changes, the rise time will be increased by the analyzer deadtime if the controller gain is less than the inverse of the open loop gain.

Exceptions: Closed loop control will not work for erratic analyzer values or poor signal-to-noise ratios.

Insight: An enhanced PID can simplify the tuning and improve the stability for loops with large and/or variable analyzer update times.

Rule-of-Thumb: Use an enhanced PID with feedback plus feedforward control for at-line and off-line analyzers.

Tip #101: Use Plant-Wide Feedforward for More Flexible and Efficient Production[13]

Greg McMillan (Technical)

The pressure to reduce inventories, coupled with changing market demands and fluctuations in raw material and energy prices, requires a process plant to be able to respond quickly to maintain optimum operation. To most efficiently support changes in production rate or product grade, flows must change in unison throughout the plant. Feedback loops cannot do this. Feedback loops will eventually arrive at the right flows but time will be lost and product quality may suffer in the interim. Feedforward control allows the plant automation system to respond immediately *in unison* to changes in production rate or product grade. During normal production, feedforward control can also compensate for composition, pressure, and temperature upsets before they have a significant impact on production rates or product quality.

Plant-wide feedforward can preemptively move a plant to match the flows and conditions (e.g., temperature and composition) of all important process streams on a process flow diagram (PFD) for a given product and production rate, thereby maintaining material, component, charge, and energy balances. Plant-wide feedforward flow control provides the fastest and least disruptive transition to new operating conditions.

Traditionally, feedforward has been implemented on just the most important loops in the most critical unit operations. The benefit is localized and the other loops are left to fend for themselves. The result is that the plant does not fully reach the new production rate or product grade until the changes in process conditions propagate from one end of the plant to the other. The plant does not settle out until each of the unit operations settles out. For feedback control, the settling time of a loop is at least four deadtimes, based on tuning practices. The total settling time can be roughly estimated as the sum of the settling times of the loops in series. In contrast, plant-wide feedforward control can reduce the plant settling time to the deadtime and settling time of the slowest loop for perfect and imperfect feedforward, respectively. Perfect feedforward requires no feedback correction.

Feedforward control has the ability to provide the preemptive correction of any disturbance that can be measured. The disturbance can be due to maintenance, abnormal conditions, operator actions or sequences, or to changes in load (feed), raw materials, recycle streams, utilities, and/or environmental conditions. The source of sequences can be the batch manager, sequential function charts, and safety instrumentation systems. In feedforward control, a disturbance is measured, filtered, multiplied by a feedforward gain, passed through deadtime and lead-lag blocks for dynamic compensation, and used as the PID output with the proper direction (reverse or direct).

To get the maximum benefit from feedforward, the correction must arrive at the same point in the process and at the same time as the disturbance, and with an effect equal but opposite to the disturbance. If the feedforward signal arrives too soon, the initial process response will be in the opposite direction of the final response to the disturbance. The result will be an inverse response. Feedback action will try to compensate for the feedforward correction, resulting in a second peak. If the feedforward arrives too late, a second disturbance will be created by feedforward action. In either case, an oscillation will develop and the error from the disturbance will be increased.

Concept: Feedforward is the most productive of the advanced PID process control techniques. Feedforward control becomes more important when production rates or grades are frequently varied to meet changing market demands or to take advantage of changes in energy costs.

Details: To avoid inflicting disturbances on a loop and other loops, filter the feedforward signal so that noise does not cause the PID output fluctuations to be larger than the control valve or variable frequency drive deadband. Set the feedforward gain to ensure that the magnitude of the feedforward signal is correct. The feedforward gain must provide a corrective change in PID output equivalent to the effect of the disturbance. Pay attention to engineering units and scales when computing the feedforward gain. To prevent a disturbance from arriving early, set a deadtime in a deadtime block that is the deadtime in the disturbance path minus the deadtime in the correction path. Note that the paths must be to the same point in the process. If the disturbance arrives before the corrective signal due to excessive deadtime in the correction path, nothing can be done via dynamic compensation to make the feedforward signal arrive sooner. There is no function for undoing deadtime. If the lag in the feedforward path is smaller than the lag in the disturbance path, add a lag to the feedforward dynamic compensation. If the lag in the feedforward path is larger than the lag in the disturbance path, add a lead time to the feedforward dynamic compensation. Use a feedforward summer and provide a visible and adjustable ratio for flow feedforward (Tip #93). If it can be done while meeting production goals, ramp production rates down as quickly as possible when energy costs are high (e.g., the cost of electricity and cooling water during the heat of the day). Maximize production rates when utility costs are low. For parallel trains of unit operations such as crystallizers and heat exchangers, maximize the feed to the more efficient units as decided online by inferential measurements of overall heat transfer coefficients (UA). For continuous crystallizers, use an inferential measurement of UA to determine when a crystallizer should be defrosted and the flow divided among the remaining crystallizers. Use a similar strategy for catalyst or ionic bed regeneration. When units are started up, bring units to operating conditions based on feedforward ratio control before switching to feedback control.

Watch-outs: When the feedforward signal goes to a valve, the feedforward gain is inversely proportional to the slope of the nonlinear installed valve characteristic; that is, the valve gain. Signal characterization in the DCS can be used to compensate for the nonlinearity (Tip #84) but the installed characteristic is often not known exactly and varies with pressure and frictional losses.

Exceptions: If the feedforward deadtime is larger than the loop deadtime, feedforward will do more harm than good and should not be used until changes in the final control element or secondary process can be made to decrease the feedforward correction path deadtime or increase the disturbance path deadtime. The disturbance often originates from another loop (such as level) or from an operator-initiated change. A deadtime in the operator initiated setpoint change can be inserted to delay the entry of the disturbance into the process. The disadvantage of the delayed action of a level loop or operator change is often less important than the disadvantage of a feedforward arriving too late.

Insight: The use of plant-wide feedforward control can move a plant quickly and smoothly for more flexible and efficient manufacturing.

Rule of Thumb: Install flowmeters and implement feedforward control so all key loops move in unison to changes in production rate or product grade to meet market demands and take advantage of changing energy costs.

Appendix A: References

Articles, Books, and Papers

1. McMillan, G. "PID Tuning Rules" Online appendices, *InTech*, Jan.-Feb. 2012.
2. McMillan, G. *Good Tuning: A Pocket Guide, 3rd Edition* ISA, 2012.
3. McMillan, G. "Full Throttle Batch and Startup Response" *Control*, May 2006.
4. Shinskey, G. "The Power of External-Reset Feedback" *Control*, May 2006.
5. McMillan, G. "What Is the Key PID Feature for Basic and Advanced Control?" *Control Talk* blog, *Control*, Apr. 26, 2012.
6. McMillan, G. and Torres, H. "Effective Use of Key PID Features" Paper presented at ISA Automation Week, Orlando, 2012.
7. McMillan, G. "Don't Overlook PID in APC" *Control*, Nov. 2011.
8. Yu, C. and Luyben, W. "Analysis of Valve-Position Control for a Dual Input Process" *Ind. Eng. Chem. Fund.*, pp. 344–350, Vol. 25, No. 3, 1986.
9. Harriott, P. *Process Control* McGraw-Hill, 1964.
10. Bohl, A. and McAvoy, T. "Linear Feedback vs. Time Optimal Control. II. The Regulator Problems", *Ind. Eng. Chem.*, Process Des. Dev., Vol. 15, No. 1, 1976.
11. McMillan, G. *Essentials of Modern Measurements and Final Elements in the Process Industries*, ISA, 2010.
12. McMillan, G. "Advances in Flow and Level Measurements Enhance Process Knowledge, Control" *InTech*, Jan./Feb. 2010.
13. McMillan, G. "Feedforward Control Enables Flexible, Sustainable Manufacturing" *InTech*, Mar.-Apr. 2011.
14. McMillan, G. "Key Design Components of Final Control Elements" *InTech*, Mar.-Apr. 2010.
15. McMillan, G. "What's Your Flow Control Valve Telling You?" *Control Design*, May 2004.
16. McMillan, G. "Improve Control Loop Performance" *Chemical Processing*, Oct. 2007.
17. Shinskey, G. "Flow and Pressure Control Using Variable-Speed Motors" *Proceedings of* Control *Conference*, pp. 161-167, Chicago, 1980.
18. Al-Khalifah, M. and McMillan, G. "Control Valves versus Variable Speed Drive for Flow Control" Paper presented at ISA Automation Week, Orlando, 2012.
19. Smith, C. "Watch Out with Variable Speed Pumping" *Chemical Processing*, May 2008.
20. McMillan, G. "Unlocking the Secret Profiles of Batch Reactors" *Control*, Jul. 2008.
21. McMillan, G. "Is Wireless Process Control Ready for Prime Time" *Control*, May 2009.
22. McMillan, G. "Wireless - Overcoming Challenges of PID Control, Analyzer Applications" *InTech*, Jul.-Aug. 2010.
23. McMillan, G. "Temperature Measurement, the Control Key to Plant Performance" *InTech*, Nov.-Dec. 2010.

24. McMillan, G. *Advanced Temperature Measurement and Control* ISA, 2011.
25. McMillan, G. *Advanced pH Measurement and Control* ISA, 2004.
26. Tolliver, T. "Distillation Control" Paper presented at ISA Automation Week, Mobile, 2011.
27. McMillan, G. *Centrifugal and Axial Compressor Control* Momentum Press, 2010.
28. McMillan, G. *Process Control Case Histories* Momentum Press, 2010.
29. Trevathan, V. *A Guide to the Automation Book of Knowledge* ISA, 2006.
30. Blevins, T. et al *Advanced Control Unleashed* ISA, 2003.
31. McMillan, G. "Wireless Measurement and Control Opportunities" Paper presented at ISA Automation Week, Mobile, 2011.
32. McMillan, G. "pH Measurement and Control Opportunities" Paper presented at ISA Automation Week, Mobile, 2011.
33. McMillan, G. "Reactor Control Opportunities" Paper presented at ISA Automation Week, Mobile, 2011.
34. McMillan, G. "Corn Analysis Modeling and Control for Ethanol Yield Improvement" Paper presented at ISA Automation Week, Mobile, 2011.
35. McMillan, G., Reedy, R. and Moulis, J. "Virtual Plant Provides Real Insights" *Chemical Processing*, 2009.
36. Moore, R. "Good pH Measurements in Bad Process Streams" *Instruments and Control Systems*, Dec. 1990.
37. McMillan, G. and Sowell M. "Virtual Control of Real pH" *Control*, Nov, 2007
38. Luyben, W. and Melcic, M. "Consider Reactor Control Lags" *Hydrocarbon Processing*, March, 1978
39. Boudreau, M. and McMillan, G. *New Directions in Bioprocess Modeling and Control* ISA, 2006
40. McMillan, G. "Optimize Batch Capacity and Efficiency" *Control*, September, 2012

For more advanced tips on PID control, see the *Control Talk* Column series with high end user Mark Congiundi that starts with the *Control* May 2011 column "35 Years of Extraordinary PID Innovations"

Website Links to Blogs and Columns

Control Talk Blog: http://community.controlglobal.com/controltalkblog

Control Talk Column: http://www.controlglobal.com/voices/mcmillan_weiner.html

ISA Interchange Blog: http://automation.isa.org/author/gregmcmillan/

Modeling and Control Blog: http://modelingandcontrol.com/author/Greg-McMillan/

Appendix B: Vintage Books

1. Bates, Roger G. *Determination of pH - Theory and Practice* John Wiley & Sons, 1973, ISBN 0-471-05647-2.
2. Franks, Roger, G.E. *Modeling and Simulation in Chemical Engineering* Wiley-Interscience, 1972, ISBN 0-471-27535-2.
3. Galster, Helmuth *pH Measurement: fundamentals, methods, applications, instrumentation* (Translated by Frank Hampson), VCH publishers, 1991, ISBN 3-527-28237-8.
4. Harriott, Peter *Process Control* McGraw-Hill, 1964, ISBN 07-026785-5.
5. Himmelblau, David M. *Basic Principles and Calculations in Chemical Engineering* 4th edition Prentice-Hall, 1982, ISBN 0-13-066498-7.
6. Luyben, William L. *Practical Distillation Control* Van Nostrand Reinhold, 1992, ISBN 0-442-00601-2.
7. Luyben, William L. and Luyben, Michael L. *Essentials of Process Control* McGraw-Hill, 1997, ISBN 0-07-039172-6.
8. Peters, Max C. *Elementary Chemical Engineering* McGraw-Hill, 1954, Library of Congress Catalog Card Number 54-7358.
9. Shinskey, F. G. *pH and pION Control in Process and Waste Streams* John Wiley & Sons, 1973, ISBN 0-471-78640-3.
10. Shinskey, F. G. *Energy Conservation Through Control* Academic Press, 1978, ISBN 0-12-641650-8.
11. Shinskey, F. G. *Controlling Multivariable Processes* ISA, 1981, ISBN 0-87664-529-5.
12. Shinskey, F. G. *Distillation Control - For Productivity and Energy Conservation* 2nd edition, McGraw-Hill, 1984, ISBN 0-07-56894-4.
13. Shinskey, F. G. *Feedback Controllers for the Process Industries* McGraw-Hill, 1994, ISBN 0-07-056905-3.
14. Westcott, Clark C. *pH Measurements* Academic Press, 1978, ISBN 0-12-745150-1.
15. Shinskey, F. G. *Process Control Systems* 4th edition, McGraw-Hill, 1996, ISBN 0-07-057101-5.

Appendix C: Checklists

All of the checklists assume that the user has made sure all of the automation system components and wiring connections meet electrical area classifications and standards and plant practices.

1. Checklist for Best Control Valve Performance

The following checklist is not intended to cover all the specification requirements, but instead some of the major application details to be addressed for this automation component. The following list assumes that the materials of construction have been properly specified and that the valve will work safely and reliably and with acceptable rangeability for the maximum possible temperature and minimum available pressure drop. Chapters 7 and 8 in the ISA book *Essentials of Modern Measurements and Final Elements in the Process Industries* document these and other considerations for maximizing the performance of control valves for process control (reference 11 in Appendix A). Reliability, precision (backlash and stiction), and rangeability are most important.

1. Does valve sizing software include the fluid physical properties for worst case operating conditions?
2. Do location and valve type eliminate or reduce damage from flashing and erosion?
3. Did you include piping reducing factor that decreases the effective flow coefficient?
4. Did you compute and plot the installed valve characteristic for worst case operating conditions?
5. Is the actuator sized to deliver 150% of the maximum torque or thrust required?
6. Is actuator threshold sensitivity better than 0.1%?
7. Is positioner threshold sensitivity better than 0.1%?
8. Is the smart positioner tuned for the application (otherwise you have a dumb positioner)?
9. Is total valve assembly deadband less than 0.4% over the entire throttle range?
10. Is total valve assembly resolution better than 0.2% over the entire throttle range?
11. Is the installed characteristic slope > 0.5% max. flow per % signal over the entire throttle range?
12. Is the installed characteristic slope < 2.0% max. flow per % signal over the entire throttle range?

Note that installed valve characteristic slope is valve gain when engineering units are used for flow rate and the split range amplification effect is included. For a 50% split range point of two control valves, the split range amplification is a factor of 2. For small and large valves installed at the same control point, the effect of size on valve gain can be mitigated by intelligent selection of the split range point. For example, for parallel valves on the same process fluid where the large valve has four times the capacity of the small valve, the split range would 0–20% for the small valve and 20–100% for the big valve. For valves on different process streams, the process gain must be included in the calculation of the intelligent split range point to reduce the open loop gain nonlinearity. Because Operations is accustomed to a 50% split range, special operator interface graphics and training are necessary.

The threshold sensitivity and resolution are the smallest input change that will cause an automation output to respond. For resolution the response is a step the size of the resolution limit (stair-step or quantized response). For threshold sensitivity, once the response occurs the output change matches the input change. The response of actuators and positioners is commonly characterized by threshold sensitivity, whereas the response of a valve with stiction (e.g., stick-slip) is characterized by resolution, assuming that the slip equals the stick. Threshold sensitivity and resolution will cause a limit cycle for a controller in automatic mode if there is integrating action in the process or in the controller via the integral mode. Deadband will cause a limit cycle if there are more than two occurrences of integrating action (e.g., an integrating process such as level and integrating action in a level controller or in a cascade control loop where integrating action in used in both the secondary and primary controllers). For a more detailed discussion of limit cycles, including the effect of tuning, see the *Control Design* May 2004 article "What's Your Control Valve Telling You?".

2. Checklist for Inline Flowmeters

The straight run requirements for inline flowmeters (Coriolis, magnetic, turbine, and vortex meters) have not been detailed to the same extent as those for orifice meters, flow tubes, and venturi differential head flowmeters. The supplier is typically the source of most of the expertise. In an attempt to provide some guidance, I have roughly correlated straight run requirements for vortex, turbine, and magnetic flowmeters to the ASME guideline for flow tubes. These ball park guidelines should be superseded by supplier, research, and user information and publications that quantify piping system effects on inline flowmeter accuracy, noise, and rangeability.

The following checklist is not intended to cover all the specification requirements, but instead some of the major application-related details to be addressed for the selection and installation of inline flowmeters. It assumes that the materials of construction have been properly specified, and the meter will work safely and reliably with acceptable accuracy at the maximum possible temperature. O-rings and gaskets are added to the checklist because potential degradation from chemical attack and temperature is often overlooked. For more information on flow measurements see the March 2012 Control Talk column "Going with the Flow." For a detailed understanding, see Chapter 4 in the ISA book *Essentials of Modern Measurements and Final Elements in the Process Industries* (reference 11 in Appendix A).

1. Do the meter's threshold sensitivity, repeatability, and drift meet application requirements?
2. Do the meter's rangeability and permanent pressure loss meet application requirements? (Maximum possible rangeability: 15:1 vortex, 50:1 turbine, 100:1 magmeter, 200:1 Coriolis.)
3. Do O-rings and gaskets meet worst case corrosive and temperature operating conditions?

4. Are gaskets projecting into the flow stream?
5. Is the meter centerline concentric with the piping centerline?
6. For vortex meters, do the upstream and downstream straight run lengths meet the ASME guideline for 0.8 beta ratio flow tubes (e.g., 20 pipe diameters upstream for long bends)?
7. For turbine meters, do the upstream and downstream straight run lengths meet the ASME guideline for 0.6 beta ratio flow tubes (e.g., 10 pipe diameters upstream for long bends)?
8. For magnetic meters, do the upstream and downstream straight run lengths meet the ASME guideline for 0.4 beta ratio flow tubes (e.g., 5 pipe diameters upstream for long bends)?
9. For turbine and vortex meters, have asymmetric flow profiles and swirling been minimized by piping design and straightening vanes or special flow conditioners for profile distortion?
10. Is the maximum kinematic viscosity less than the maximum permissible for vortex meters?
11. For magnetic, turbine, and vortex meters, are the minimum and maximum velocity within limits?
12. For vortex meters, is the minimum Reynolds number greater than the minimum required by the meter?
13. Are flowmeters in vertical lines installed with flow up?
14. For lined magnetic meters, is maximum vacuum (e.g., after steam cleaning) less than the maximum the meter can withstand?
15. For magnetic meters, is the minimum fluid conductivity greater than the minimum detectable by the meter?
16. Are there/will there be bubbles in magnetic, turbine, and vortex meter applications?
17. Are the maximum % bubbles and solids less than the maximum permitted by Coriolis meter software?
18. For U-tube Coriolis, magnetic, turbine, and vortex meters, is particle abrasion negligible?
19. Is particle concentration high enough to require a straight tube Coriolis meter?
20. For turbine meter bearings, is the minimum fluid lubricating effect better than the minimum required by the meter?
21. For magnetic meters, is the fluid always a liquid (e.g., no flashing)?
22. Are Coriolis and magnetic flowmeters completely full at zero flow?
23. To prevent sloshing errors, is the signal grounded to zero when flow is zero?
24. Is maximum piping vibration less than the maximum permitted by Coriolis and vortex meters (e.g., is there a vibration damper for isolation if required)?
25. Are bubbles and solids being trapped in U-tube Coriolis, magnetic, turbine, and vortex meters?
26. Are magnetic meters properly grounded to earth and for lined pipe, are there ground straps between pipe flanges and the meter?

3. Checklist for Loop Analysis by Trend Charts

If historian data compression is too great, disturbances, limit cycles, and noise will not be visible in a trend chart. To be able to properly assess loop performance, the compression of the process variable (PV) should be less than 1/5 of the control band, where the control band is the allowable PV error around setpoint (SP). To detect limit cycles, the compression of the controller output (OUT) must be less than the control valve or variable frequency drive deadband, resolution, and threshold sensitivity. The limit cycles in the OUT tend to be "sawtooth" in shape. The limit cycles in flow or liquid pressure tend to be square wave. Limit cycles in level and temperature tend to be sawtooth. For controller gains greater than 1, patterns are more apparent in the OUT or the manipulated flow. To analyze noise, the compression of the PV must be less than 1/5 of the noise amplitude. To monitor changes in noise that are symptomatic of changes in the sensor or process, the unfiltered PV should be plotted. A reduction in noise may indicate a coated thermowell or electrode. An increase in noise may indicate bubbles in gas streams or droplets in vapor streams, or a decrease in the degree of mixing.

If the time span is too short, load and setpoint responses, slow rolling oscillations in integrating processes, and limit cycles will not be evident and the source of a disturbance will not be detectable. The response will look like a sloping horizontal line. If the time span is too long, the trend will look like a squished jagged series of spikes around setpoint. Patterns and what started first will not be recognizable. To detect limit cycles from backlash and stiction, the time span should be about 20 times the integral time. To detect slow rolling oscillations in integrating processes from too low of a controller gain or too low of a reset time, the time span should be about 100 times the integral time. Equation C-14a in the appendices for the *InTech* Jan.-Feb. 2012 article "PID Tuning Rules" (reference 1 in Appendix A) shows the relationship that triggers these oscillations. To determine which flow caused an upset and see the PID response, the time span should be about 20 deadtimes. To identify a temperature or composition disturbance, the time span should be about 100 deadtimes.

Because nearly all process inputs are flows, the first flow to change in a unit operation is most likely the source of the upset to the unit operation. Most upsets start out as flow changes from poor valves, poor controller tuning, batch operations, manual actions, on-off actions, sequences, or trips. Changes in temperature or composition are much slower and more difficult to detect. The upset will be seen in the OUT or flow manipulated by a loop affected by the change.

To see the setpoint response including the settling time, the time span should be greater than 10 times the rise time. The rise time is the deadtime plus the setpoint change divided by the PV rate of change. The rate of change (ramp rate) is best measured online but can be estimated as the near-integrating process gain multiplied by the change in controller output from the setpoint change. The future PV value plus the PV ramp rate (described in the Jul. 29, 2012 Control Talk blog the "Future PV Values Are the Future") should be on the trend plot with the loop PV, SP, and OUT for setpoint response analysis.

The following checklist does not cover all of the potential uses of trend charts but will get you started with the setup of effective trend charts:

1. Is the historian PV compression less than 1/5 the control band and noise amplitude?

2. To see limit cycles, is the historian OUT compression less than the control valve or VFD deadband, resolution, or threshold sensitivity?
3. To be able to estimate loop deadtime is the historian sample time less than 1/5 the deadtime?
4. To see an oscillation amplitude, is the historian sample time less than 1/10 the period?
5. For noise analysis, is the unfiltered PV displayed on a trend chart?
6. For disturbance analysis, are all process and utility flows for a unit operation, besides loop PV, SP, and OUT, on the same chart?
7. For setpoint response analysis, is the PV ramp rate and future PV besides loop PV, SP, and OUT (Tips #89 and 90) on the same chart?
8. For analyzing flow disturbances, is the time span about 20 deadtimes?
9. Is the time span about 20 times the reset time to see limit cycles?
10. Is the time span about 100 deadtimes to analyze temperature disturbances?
11. Is the time span about 100 times the reset time to see slow rolling oscillation in integrating processes?
12. Are the integrated error and peak error for load disturbances and the rise time, overshoot, undershoot, and settling time for setpoint changes on a trend chart with the loop PV, SP, and OUT?

4. Checklist for pH Measurement

The composition measurement with the greatest sensitivity and rangeability by far is pH. Consider that pH measurement routinely detects changes at 7 pH to the 9th decimal place and for a 0-14 pH scale covers 14 orders of magnitude of hydrogen ion concentration. New glass and reference designs have dramatically reduced drift even for the harshest conditions, such as high temperature and sterilization. Reference junctions can now be easily replaced. Smart wireless transmitters have much better diagnostics and resolution. Here is a checklist to help you take advantage of these advances in pH measurement technology.

This checklist is not intended to cover all the specification requirements, but some of the major application details to be addressed for this automation component. The following list assumes that the materials of construction have been properly specified and that the sensor will work safely and reliably and with acceptable accuracy for the maximum possible temperature and pressure. O-rings and gaskets are added to the checklist because potential degradation from chemical attack and temperature is often overlooked. The slides referenced in parentheses below are in the ISA Automation Week 2011 tutorial "pH Measurement and Control Opportunities" (reference 32 in Appendix A). For a greater understanding of pH measurement, see Chapter 6 in the ISA book, *Essentials of Modern Measurements and Final Elements in the Process Industry* (reference 11 in Appendix A).

1. Do O-rings and gaskets meet the requirements of worst-case corrosive and temperature operating conditions?
2. Is the best glass used for the worst-case temperature, pH, and chemicals that can attack glass? (e.g., general purpose, high pH, high temperature, sterilizable, HF resistant)?
3. Is a spherical bulb used for maximum accuracy?

4. For pH < 1 or > 12 would conductivity or density give a better concentration measurement?
5. Are the best reference design and fill used for the accuracy and speed requirement and worst case temperature and composition? (Examples of worst case composition include low water or pure water solutions that have low conductivity and salt and chemical concentrations that change junction potential, plug junctions, and poison reference internals.)
6. Is the accuracy and equilibration speed requirement or the coating and poisoning problem so extraordinary that a flowing junction is needed to provide the most constant reference potential and the fastest junction equilibration, and eliminate plugging and/or poisoning?
7. If plugging is not a problem, can an aperture junction be used to give the lowest junction potential?
8. Will double and triple junction references be sufficient to slow down the contamination rate?
9. Are special electrolytes needed to prevent precipitation of silver salt from contact with a process component (e.g., silver cyanide precipitate from process cyanide contact with silver)?
10. Can a removable reference junction enable an electrode to be rejuvenated (reference junction and fill replaced) before plugging /poisoning becomes a problem (slide 26)?
11. Can a large surface solid reference electrode essentially eliminate plugging, contamination, and poisoning if reference speed of equilibration is not a problem (slide 19)?
12. Is chemical attack, premature aging from high temperature, or dehydration (non-aqueous solvents or low water concentrations) so severe that an automated retractable insertion assembly is needed to limit sensor exposure to the process just long enough to get periodic pH measurement?
13. Is the solution conductivity so low (e.g., condensate, boiler feedwater, deionized water) that a special assembly is needed to provide low sample flow, diffuser, and electrolyte reservoir (slide 22)?
14. Can a Varipol (VP) or equivalent connector be used to quickly locally disconnect the electrode cable, eliminating the need to disconnect the transmitter and retract the cable through conduit or flex to prevent twisting the cable on disconnection?
15. Is a smart electrode with a stored calibration record available for the selected electrode (slide 40)?
16. Is a solution ground needed for impedance diagnostics and ground potential elimination?
17. Is a smart transmitter available to detect glass and reference problems (slides 27-32)?
18. Is solution pH temperature compensation needed in addition to standard Nernst temperature compensation (slides 6, 29)?
19. Can a wireless transmitter be used to get the latest features and enable portability of the measurement to test the best electrode and location (least deadtime and least noise/bubbles) (slide 42)?
20. Is the electrode installed with the tip pointing down at a 30-60° angle to prevent bubbles from residing in the tip or being caught on the internal electrode (slide 33)?
21. Are electrodes always wetted, even for batching and during shutdown of continuous operations?

22. Is middle signal selection needed to eliminate response to single failure and noise (slide 16)?
23. Are stream velocity and protective shroud design the best for the process conditions (slides 17, 18)?
24. Are a stream velocity of 5–10 fps and exposure of the glass to flow needed to prevent coating?
25. Are a stream velocity of 0.1–1 fps and shroud reducing flow impingement needed to decrease damage from abrasion?
26. Does the electrode tip extend into the center line of a pipe and past the baffles in a vessel?
27. Is the electrode location free from bubbles (e.g., not near the sparger ring)?
28. Are the electrodes sufficiently downstream from a pump or static mixer to reduce concentration and pressure fluctuations but not so far as to increase deadtime by more than 3 sec (slides 33, 34)?
29. Are insertion electrodes in series so each electrodes is at the same velocity and sees the same composition (slides 33, 34)?
30. Is electrode location free from flashing (e.g. not on pump suction or valve discharge)?
31. Should an insertion type of electrode assembly with a ball valve and restraining strap be used to be able to safely withdrawn (retract) the electrode from a pipe or vessel?
32. Is the electrode and transmitter location safely accessible for maintenance?
33. Are the electrode and transmitter signal connections always dry?

5. Checklist for PID Controller Features

There is an incredibly broad offering of PID features and options, enabling maximum disturbance rejection and setpoint response but also the coordination of loops and unit operation optimization. While the full range of modern PID controller capability is book-worthy, this checklist and an excerpt from an ISA Automation Week 2012 paper can get you started on the right path.

If you don't get the valve action and control action right, nothing else matters. The controller output will ramp off to an output limit. The valve action (inc-open and inc-close) can be set in many different places, such as the PID block, analog output (AO) block, splitter block, signal characterizer block, current to pneumatic (I/P) transducer, or positioner. Make sure the valve signal is not reversed in more than one location for an inc-close (fail open) valve. Once the valve action is set properly, set the control action to be the opposite of the process action. The control action is reverse or direct if an increase in the PID output causes the PID process variable (PV) to increase or decrease, respectively. Verify with the process engineer the valve action, process action, and resulting control action required.

The following checklist is not intended to cover all the configuration requirements, but some of the major application details to be addressed for PID controllers. Most of the power of the modern PID controller is untapped. For more information see the ISA Automation Week 2012 paper "Effective Use of PID Features," appendices for the *InTech* Jan.-Feb. 2012 article "PID Tuning Rules," and the ISA book *Good Tuning: A Pocket Guide* - 3rd edition (references 1 and 2 in Appendix A).

1. Does the measurement scale cover the entire operating range, including abnormal conditions?
2. Is the valve action correct (inc-open for fail close and inc-close for fail open)?
3. Is the control action correct (direct for reverse process and reverse for direct process if the valve action is set)?
4. Is the best "Form" selected (ISA standard form)?
5. Is the "obey setpoint limits in cascade and remote cascade mode" option selected?
6. Are the "back calculate" signals correctly connected between blocks for bumpless transfer?
7. Is the "PV for back calculate" selected in the secondary loop PID?
8. Is the best "Structure" selected (PI action on error, D action on PV for most loops)?
9. Is the "setpoint track PV in manual" option selected to provide a faster initial setpoint response unless the setpoint must be saved in PID?
10. Are setpoint limits set to match process, equipment, and valve constraints?
11. Are output limits set to match process, equipment, and valve constraints?
12. Are anti-reset windup (ARW) limits set to match output limits?
13. Is the module scan rate (PID execution time) less than 10% of minimum reset time?
14. Is the signal filter time less than 10% of minimum reset time?
15. Is the PID tuned with a proven tuning method or by an auto-tuner or adaptive tuner?
16. Is the rate time less than ½ the deadtime (the rate is typically zero except for temperature loops)?
17. Is external-reset feedback (dynamic reset limit) enabled for cascade control, analog output (AO) setpoint rate limits, and slow control valves or variable speed drives?
18. Are AO setpoint rate limits set for blending, valve position control, and surge valves?
19. Is integral deadband greater than limit cycle PV amplitude from deadband and resolution?
20. Can an enhanced PID be used for loops with wireless instruments or analyzers?

6. Checklist for Pressure and Differential Pressure (DP) Measurements

Pressure and DP measurements are important for regulating inventory, flow, and product quality and are critical for ensuring safe process operation. Installation-induced errors and the drift of today's smart pressure and DP transmitters have decreased by an order of magnitude compared to the transmitters prior to the 1990s by the compensation of temperature, static pressure, and non-ideal sensor effects. Most of the remaining concerns are associated with impulse lines.

Pressure control of boilers, prime movers (e.g., pumps, fans, and compressors), and headers is important for supply to consistently meet demand. Pressure control in reactors, columns, evaporators, and extruders is the key to reducing variability in product quality. Pressure control is critical for increasing safety and decreasing emissions by preventing excursions that would activate safety instrument systems (SIS) and relief devices. Pressure measurements for SIS actions must be fast, reliable, and accurate.

The driving force for flow across system resistances and control valves are pressures. DP measurements can monitor the performance of filters and the coating of heat exchanger surfaces. Portable wireless DP transmitters can generate process diagnostics and identify installed valve flow characteristics. DP measurements in clean rooms can maintain a differential of 0.25 inches water column for positive air flow out of the room to prevent contamination in pharmaceutical and electronics manufacturing.

DP is by far the most prevalent type of instrument used for level and flow measurement despite newer technologies, such as radar and Coriolis, primarily due to cost. Material balance control, residence time control, flow ratio control, and vessel-sump-tank inventory control most often depend on DP measurements. High levels can cause spills and carry-over of the process fluid into vent systems. Low levels can damage pumps.

Given the significant role of pressure and DP measurements in manufacturing, what can be done to get the best implementation?

The following checklist is not intended to cover all the installation and specification requirements, but some of the major application details to be addressed for this automation component. The following list assumes that the materials of construction have been properly specified and that the sensor will work safely and reliably and with acceptable accuracy at the maximum possible temperature. For a detailed understanding see Chapters 3-5 in the ISA book *Essentials of Modern Measurements and Final Elements in the Process Industries* (reference 11 in Appendix A).

1. Is the transmitter fast enough for the application (e.g., damping < 0.2 sec for compressor control)?
2. For gas, is the transmitter mounted above the process connection to prevent the accumulation of liquids?
3. For liquid, is the transmitter below the process connection to prevent the trapping of gases?
4. Do impulse lines have a continuous slope with no bends or with smooth, long radius bends?
5. Do impulse lines have vent and drain valves?
6. Does a DP have an equalization valve?
7. Does process pressure connection design prevent adverse velocity head?
8. Do transmitter and impulse lines need freeze protection?
9. If heat tracing is used, are high temperatures prevented that could alter fluid composition in impulse or transmitter lines (e.g., vaporization, reactions, or the formation of tars and polymers)?
10. For plugging services, can impulse lines be purged or eliminated?
11. For purged impulse lines, are purge flow and pressure high enough?
12. For purged impulse lines, are purge flow and pressure indicated and adjustable?
13. For purged impulse lines, is liquid purge needed to eliminate transients resulting from compressibility of purge during fast static pressure disturbances and to prevent solids build-up at the bubbler tip?
14. Can impulse lines be eliminated by direct mounting of the transmitter or by the use of a capillary system?

15. Should an extended diaphragm be used to minimize fouling of a direct mounted level DP?
16. Does the diaphragm area need to be increased to increase threshold sensitivity?
17. Does the diaphragm area need to be decreased to increase speed of response?
18. Is capillary length minimized to increase speed of response?
19. Are all capillary systems at the same temperature (e.g., sun versus shade)?
20. For DP measurement with low static pressure, can DP be computed from dual direct-mounted transmitters to eliminate impulse lines?
21. Can a smart transmitter be used to detect plugged impulse lines?
22. Can wireless transmitters be used to provide portability for process troubleshooting?

7. Checklist for Radar Level Measurement

Radar offers a highly sensitive measurement of surface level that can be nearly maintenance-free.

Its biggest benefit is accurate accounting of raw material and product tank inventories. In addition, a rate of level change measurement, made by simply sending the level through a deadtime block with a deadtime large enough to show an appreciable change in level to maximize the signal-to-noise ratio, can provide a feed rate measurement that rivals that of a mass flowmeter for a material with a consistent composition. In many raw material and product storage tanks the composition and hence the density relationship with temperature is well defined from periodic analysis.

An extremely accurate level measurement can enable material balance and composition control and analysis as discussed in "Advances in Flow and Level Measurements Enhance Process Knowledge, Control" (reference 12 in Appendix A).The use of radar measurement on distillate receiver level can enable internal reflux control and the use of the preferred material balance control scheme when the reflux flow is high enough that distillate level manipulates reflux flow. The use of radar measurement on reactor level can enable residence time control for conversion control as seen on slides 6-9 and 10–16 in the ISA Automation Week 2011 tutorial "Reactor Control Opportunities" (reference 33 in Appendix A).

The principal consideration in the selection and application of radar is associated with the fact that radar level is a surface measurement. Vortexes, turbulence, and even just ripples in the surface can cause significant noise. A slanted surface can cause a "no return" signal and until recently a very high level could be misread as no level. Success depends upon a careful review of vessel operating conditions, mixing, and geometry. Maintenance can be minimal if changes to the process and vessel are reviewed as to their effects on the radar installation.

At extremely low levels, surface abnormalities as the vessel or tank is being emptied, possibly creating something similar to what you see in the tank of a flushing toilet, may necessitate of the installation of a different measurement technology as a backup.

The potential for accurate inventory measurement is dramatic, but with this extreme capability come some extraordinary application considerations. This checklist offers a number of implementation details to help achieve the full capability of this measurement device.

This checklist is not intended to cover all the specification requirements, but some of the major application details to be addressed for both non-contacting and guided wave radar. The following list assumes that the materials of construction have been properly specified and that the sensor will work safely and reliably and with acceptable accuracy at the maximum possible temperature. For more information see the *Chemical Processing* Jul. 2011 article "Making the Most of Radar" and the *Control* magazine Feb. 2012 Control Talk column "Radar Love." For a detailed understanding see Chapters 5 in the ISA book *Essentials of Modern Measurements and Final Elements in the Process Industries* (reference 11 in Appendix A). Reliability and precision (noise, repeatability, resolution, and threshold sensitivity) are most important.

1. Is the dielectric constant of the liquid too low for even guided wave radar?
2. Is software available to improve signal strength and ignore false echoes?
3. If foam is present, do you want to detect the foam surface, or the liquid surface?
4. Is a stilling well needed to reduce turbulence and foam?
5. Will the return signal be affected by gaps/holes in the stilling well?'
6. Will the tank bottom reflect signals, causing false returns?
7. Is the non-contacting beam or guided wave radar probe located away from the vessel center and any agitators, coils, and inlet streams?
8. Is the path open enough for non-contacting radar?
9. Is the nozzle large enough for the cone (horn) antenna preferred for non-contacting radar?
10. Will the nozzle neck be too long, interfering with the horn antenna?
11. For tall tanks and low dielectric material, is the antenna large enough to handle the range and dielectric?
12. Is the antenna size matched to stilling well size, where a stilling well is used?
13. Is high frequency radar needed for the non-contacting beam to be narrow enough for a tall tank and to avoid vessel internals?
14. Is high frequency radar needed to allow a recessed antenna or a full port valve in the nozzle?
15. Is there too much vapor, foam, or condensation for high frequency radar?
16. Will the highest level, including foam and swell, be sufficiently below the radar antenna?
17. Is the fluid too viscous, sticky, abrasive, or corrosive for guided wave radar?
18. Is the dielectric constant so low that guided wave radar is needed?
19. Is the signal-to-noise ratio so low that guided wave radar is needed?
20. Is the surface so slanted that receiving a reflected signal with a non-contact device is unlikely, requiring the use of guided wave radar?
21. Is the minimum clearance between a guided-wave probe and vessel internals greater than 4 inches?
22. Is the stilling well diameter greater than 4 inches for guided wave radar?
23. Do coatings and deposits require the use of a single lead guided wave probe?
24. Do obstructing objects require the use of coaxial guided wave probes?
25. Does a low dielectric constant require the use of coaxial guided wave probes?

26. Does a viscous non-coating fluid require twin guided wave probes?
27. Is the vessel so tall that flexible guided wave probes are needed for level measurement range?
28. Does the guided wave radar probe need to be anchored to the vessel bottom to reduce sway?
29. Does a DP need to be used for low level measurement due to an erratically shaped surface (e.g., voids and vortexes) when the vessel is nearly empty?
30. Is there any need for separate lightning arrestors on top of the tank?
31. Is the tank properly grounded to minimize noise and the transformer effect?
32. If an electronic calibration simulation is prepared for installation, will it match actual conditions?
33. Does the electronics housing allow the removal of components for repairs while in service?

8. Checklist for Temperature Measurement

Temperature is the most important common measurement for product quality. Temperature determines product formation rate and yield in reaction and crystallization processes, and product composition in separation processes. Temperature is also important for monitoring energy balances and heat transfer coefficients to reduce energy use. The principal considerations are the selection of a TC or RTD and the best thermowell assembly design, installation, and location for the application.

Integral (head) mounted wireless transmitters on thermowells offer the portability needed to demonstrate optimum sensor locations and online analysis of heat transfer coefficients and inferential measurement of heat transfer rate to optimize reaction rates and process efficiency as noted in the ISA Automation Week 2011 paper "Wireless Measurement and Control Opportunities" (reference 31 in Appendix A).

Portable transmitters can also be used to find the best tray for column temperature control (e.g., the tray with the largest and most symmetrical temperature change for a change in reflux to feed ratio). For more information on the importance of temperature for controlling processes, see the Control Talk Blogs on Mar. 28, 2012 "How Can You Quickly Increase Production Rate and Efficiency? (Part 2)" and on April 5, 2012 "How Can You Quickly Increase Production Rate and Efficiency? (Part 3)"

The following checklist is not intended to cover all the specification requirements, but some of the major application details to be addressed for resistance temperature detectors (RTD) and thermocouples (TC). The following list assumes that the materials of construction have been properly specified and the sensor will work safely and reliably and with acceptable accuracy for the maximum possible temperature. For more information see Chapters 1-2 in the ISA book *Advanced Temperature Measurement and Control.* (reference 24 in Appendix A). Reliability and precision (noise, repeatability, resolution, and threshold sensitivity) are most important.

1. Is the distance between the equipment outlet (e.g., heat exchanger exit) and sensor at least 25 pipe diameters (for a single phase flow) to promote mixing (recombination of outlet streams)?

2. Is the transportation delay (distance divided by velocity) from the equipment outlet (e.g., heat exchanger exit) to the sensor less than 5 seconds (e.g.; 50 pipe diameters of 6″ pipe at 5 fps)?
3. Does the distance from a desuperheater outlet to the first elbow provide a residence time (distance/velocity) that is greater than 0.1 sec?
4. Does the distance from a desuperheater outlet to the sensor provide a residence time (distance/velocity) that is greater than 0.2 sec?
5. If application vibration is not excessive, is an RTD used for temperatures below 400 °C to improve threshold sensitivity, drift, and repeatability by more than a factor of ten compared to a TC?
6. For RTDs operating at temperatures above 400 °C, are length minimized and sheath diameter maximized to reduce error from insulation deterioration?
7. For RTDs operating at temperatures above 600 °C, is the sensing element hermetically sealed and dehydrated to prevent an increase in platinum resistance from oxygen and hydrogen dissociation?
8. For TCs operating at temperatures above 600 °C, is decalibration error from changes in the composition of the TC minimized by choice of sheath and TC type?
9. For TCs operating at temperatures above 900 °C, is the sheath material compatible with the TC type?
10. For TCs operating above the temperature limit of sheaths, is the ceramic material with the best conductivity and design used to minimize measurement lag time?
11. For TCs operating above the temperature limit of sheaths and in contact with gaseous contaminants or in reducing conditions, are primary (outer) and secondary (inner) protection tubes designed to prevent contamination of the TC element and still provide a reasonably fast response?
12. In furnaces and kilns, do location and design minimize radiation and velocity errors?
13. Is the immersion length long enough to minimize heat conduction error (e.g., L/D > 5)?
14. Is the immersion length short enough to prevent vibration failure (e.g., L/D < 20)?
15. Is the process fluid velocity fast enough to minimize coating (e.g., > 5 fps)?
16. Is the process velocity fast enough to provide a fast response (e.g., > 0.5 fps)?
17. For pipes, is the tip near the centerline?
18. For vessels, does the tip extend sufficiently past the baffles (e.g., L/D > 5)?
19. For columns, does the tip extend sufficiently into the tray or packing (e.g. L/D > 5)?
20. For TCs, is it more important to minimize noise by using an ungrounded junction, or to minimize sensor element lag time by using a grounded junction?
21. To increase RTD reliability, are dual RTD elements used except where vibration failure is more likely due to smaller gauge?
22. To increase TC reliability, does the sensor have dual isolated junctions?
23. For maximum reliability, are three separate thermowells with middle signal selection used?
24. Does the sensor fit tightly into the thermowell to minimize measurement lag from air gap (e.g., annular clearance < 0.01 inch)?

25. Is an oil fill used that will not form tars or sludge at high temperature in a thermowell with the tip pointed down (to keep fill in the tip) to minimize measurement lag?
26. Is premium TC extension wire used to minimize measurement uncertainty?
27. Is four wire RTD lead wire used to minimize measurement uncertainty?
28. Are integral mounted temperature transmitters used for accessible locations to eliminate extension wire and lead wire errors and reduce noise?
29. Are wireless integral mounted transmitters used to provide portability of measurement for process control improvement and to reduce wiring installation and maintenance costs?
30. Are proper linearization tables used in the transmitter and calibrator?

9. Checklist for Best Variable Speed Pump Performance

Variable speed pumps can save energy when the flow rate is reduced. However, the desired turndown of the pump speed may not achievable if there is overheating at low flow rates, excessive flow sensitivity to pressure changes at low rpm, and cycling at high destination pressures. Special attention to variable frequency drive (VFD), motor, pump, and piping system design is necessary to achieve the rangeability required for achieving the energy savings. For tight process control, cascade control design, input card resolution, pickup design, and drive setup are important.

One of the worst things that can happen is if the speed is turned down so low that the pump (or fan) discharge head drops below the destination pressure, leading to a reversal of flow. This has occurred with the use of variable speed pumps for reactant feed flow, causing reverse flow of reactor catalyst into the reactant feed tanks. If something like this happens once, variable speed pumping may never again be allowed in the plant.

Overheating is the most commonly stated reason for poor rangeability. Motor and motor cooling design must be addressed, especially for high temperature applications such as induced draft fans on furnaces.

Poor speed control often stems from attempts to do speed control in the control room, poor resolution of the standard input signal card supplied by drive manufacturers, and excessive rate limiting and deadband introduced in the drive setup. Drive manufacturers have historically not understood the consequences of deadband, precision (e.g., resolution), and dynamics on process control.

The following checklist is not intended to cover all the specification requirements, but some of the major application details to be addressed. This list assumes that the prime mover (e.g., pump or fan) materials of construction have been properly specified and will work safely and reliably with acceptable pump discharge pressure for the maximum possible temperature and static head. The ISA Interchange post "Is a VFD or Valve Faster" and the ISA book *Essentials of Modern Measurements and Final Elements in the Process Industries* (reference 11 in Appendix A) discuss these and other considerations in maximizing the performance of variable frequency drives for process control. Reliability, precision, and rangeability are most important.

1. Is Pulse Width Modulation (PWM) drive used to reduce torque pulsation (cogging) at low speeds?
2. Is a totally enclosed fan cooled (TEFC) motor used with a constant speed fan or booster fan as necessary with class F insulation (inverter duty) and 1.15 service factor to prevent overheating? Is a totally enclosed water cooled (TEWC) motor needed for high temperatures to prevent overheating?
3. Is a NEMA frame B motor used to prevent a steep torque curve?
4. Is the pump sized to prevent operation on the flat part of the pump curve?
5. Is a recycle valve needed to keep the pump discharge pressure well above static head at low flow and a low speed limit needed to prevent reverse flow for highest possible destination pressure? See article "Watch Out for Variable Speed Pumping" (reference 19 in Appendix A).
6. Are signal input cards of greater than 12 bit used to improve the resolution limit of the signal to 0.05% or better?
7. Do the drive and motor have a generous amount of torque for the application so that speed rate-of-change limits in the drive setup do not prevent changes in speed that are fast enough to compensate for the fastest possible disturbance?
8. Is excessive deadband introduced into the drive setup, causing delay and limit cycling?
9. For tachometer control, does the magnetic or optical pickup provide enough pulses per revolution to meet the speed resolution requirement?
10. For tachometer control, is the speed control kept in the VFD to prevent violation of the cascade rule where the secondary speed loop should be 5 times faster than the primary flow loop?
11. To increase rangeability to 80:1, is fast cascade control of speed to torque in the VFD considered to provide closed loop slip control as described in *The Control Techniques Drives and Controls Handbook*, IEEE Power and Energy Series 35, Cambridge University Press, 2001?

10. Checklist for Virtual Plant

The following is a checklist to help trigger the right thought processes for arriving at a simulation that will meet your objectives for Systems Acceptance Testing (SAT), Operator Training Systems (OTS), and Process Control Improvement (PCI). Questions 18 through 30 are for first principle models. In general, you should start with a tieback model, increase tieback model fidelity by step response models, and develop a first principle model with valve, measurement, and process dynamics adjusted to match the step response models. Note that dead band, resolution, and threshold sensitivity will affect the open loop gain but not the deadtime in step response models.

1. Are the actual displays and trend charts used in the control room loaded and the actual configuration downloaded into the virtual plant?
2. For an SAT, does the simulation write and read to the actual I/O assignments?
3. Do you have the Process Flow Diagrams (PFD) and Piping & Instrument Diagrams (P&ID)?
4. Do you have the operating procedures and control definitions?

5. What are the simulation fidelities needed for SAT, OTS, and PCI?
6. Which are the most important loops for achieving the required fidelity?
7. Have you setup scenarios to automatically test the ability to deal with abnormal situations?
8. Can you play back the scenarios?
9. Do you have an automatic grading system of time and accuracy of solution for the scenarios?
10. Can you speed up the dynamics so scenarios are completed in less than an hour?
11. Have you retuned the controller for the new time constant to deadtime from speedup?
12. Have you provided a way of resetting compositions, levels, pressures, and temperatures?
13. For the important loops, do you have the open loop gain, deadtime, and open loop time constant identified from an adaptive tuner or rapid modeler for a step response model?
14. For the important loops, do you have two operating points to provide the biases for the controller output ($\%CV_0$) and process variable ($\%PV_0$) to enable the use of an open loop gain (K_0) based on deviation variables in a step response model as seen in equation below?
15. For the important loops, have you modeled the control or variable frequency drive deadband, delay, lag, deadtime, installed characteristic, rate limiting, resolution, and threshold sensitivity?
16. For the important loops, have you modeled the measurement delay, lag, noise, resolution, and threshold sensitivity, including wireless default update rate and analyzer cycle time?
17. For DCS simulate mode have you interfaced model outputs to AI block simulate inputs and AO block outputs to model inputs? (Interfacing PID will bypass AI filter and AO valve action.)
18. Do you have the chemical name, formula, and physical properties (e.g., molecular weight, density, vapor pressure, phase enthalpies, and boiling point for liquids) for each component?
19. Do you have the stoichiometric equations, yield, and kinetic equations for reactions?
20. Do you have the seed, growth, and attrition kinetic equations for crystals and cells?
21. Do you have equipment volumes and pump and compressor curves?
22. Do you have the cross-sectional areas for levels?
23. Can you use an existing library of advanced modeling objects for unit operations, piping, final control elements (e.g., control valves and variable speed drives) and measurements?
24. Do you need to write modeling objects for missing unit operations in CALC blocks using:
    a) differential equations for material, energy, and component balances for all phases as exemplified in Appendix D?
    b) charge balance equation of acids and bases for pH with dissolved carbon dioxide added as a moderator of slope between 4 and 8 pH?
    c) mixing and injection delays?

d) driving force equations for heat and mass transfer with heat and mass transfer coefficients and areas and equilibrium relationships between vapors and liquids and between gases and dissolved gases?
e) equation of state for gas pressure (e.g., ideal gas law with compressibility factor)?
f) equations for final control element and measurement dynamics?
g) equations for kinetics and population balances?
h) equations for momentum balance for compressor surge control?
i) speedup factors for differential equations and kinetic equations?

25. Have you added flow control loops as necessary to reduce the dependence on a pressure-flow solver for flows in the virtual plant to match the flows in the actual plant?
26. If the controls cannot be speeded up in unison with the process model, have you scaled flows in proportion to kinetic rate and mass transfer rate speedup, kept the deadtime about the same, and decreased the controller gain by the differential equation speedup factor?
27. Have you commissioned and tuned level and pressure loops to keep inventories in bounds?
28. Have you commissioned and tuned remaining loops with setpoints to match the PFD?
29. Have you adjusted model parameters to match manipulated flows on the PFD or use model predictive control (MPC) to automatically adapt parameters to make manipulated flows in a virtual plant running in sync with same setpoints as real plant, as described in the *Control* magazine Nov. 2007 article "Virtual Control of Real pH" (reference 37 in Appendix A)?
30. Have you adjusted model parameters to match process gains (e.g., slope of pH titration curve) as described in the *Chemical Processing* magazine article "Virtual Plant Provides Real Insights" (reference 35 in Appendix A)?

Equation (3) in Tip #89 step response model in terms of deviation variables solved for *%PV*:

$$\%PV = K_o * (\%CO - \%CO_o) + \%PV_o$$

Appendix D: Dynamics

The following is an excerpt from an ISA book on reactor control available September 2013. Key insights are in bold type. While the description of dynamics was primarily written for reactors, the dynamics are generalized to be applicable to other industrial processes.

The three types of process responses encountered in industry are defined based on an open loop test in which the PID controller is in manual or remote output so there is no response of the PID to the process (no closed loop response). To run the test, a step change is made in the PID output. The process response is observed until the type of process can be identified. During the test, there should be no disturbances, so that the process response observed is entirely the result of the step change in PID output.

The three types of processes are self-regulating, integrating, and runaway. A self-regulating process will ramp, then decelerate (i.e., rate of change continually decreases) to a new steady-state operating point (Figure D-1). An integrating process will continually ramp (Figure D-2). A runaway process will accelerate (i.e., rate of change continually increases) until it hits a relief or interlock setting (Figure D-3). For estimating loop performance and tuning settings, the parameters used to identify each type of process response are gain, time constant, and deadtime. The definition of the parameters depends upon the type of response. The parameters have alternate names in industry: for example, "lag" may be used for time constant, "delay" may be used for deadtime, and "sensitivity" may be used for gain.

The response observed in these tests includes the response of the analog output, final control element (e.g., control valve or variable frequency drive), process, sensor, transmitter, analog input, and process variable (PV) input to the PID. The observed response includes the effect of velocity limits, deadtimes, time constants, and gains in the automation system. Better terminology would be "open loop response" instead of "process response" because the observed response includes almost everything in the loop response. In addition, the source of the individual parameters that create the particular dynamic in the response should precede the term (e.g., valve deadtime or measurement deadtime).

All processes have a deadtime that is the time interval between a step change in control output and the first recognizable change in the process. Noise can delay the recognition until the excursion is beyond the noise band, creating a longer deadtime. The observed deadtime is frequently called the process deadtime. The observed deadtime is really a total loop deadtime (θ_0) that is the sum of all the pure deadtimes and the equivalent deadtimes from all time constants smaller than the largest time constant in the loop for a first order (one time constant)-plus-deadtime approximation. For a second order-plus-deadtime approximation that includes a secondary time constant, all time constants smaller than the largest time constant create an equivalent deadtime. The secondary time constant creates a bend instead of a sharp corner in the initial response right after the deadtime. The equivalent deadtime increases from 30% to 99% of the time constant as the ratio of the time constant to the largest time constant gets smaller. Time constants that are small compared to the largest time constant are summed as being essentially 100% deadtime. The ultimate limits of the peak and integrated errors are proportional to the deadtime and deadtime squared, respectively (Tip #70). **If there were no deadtime and no noise, perfect control would be possible.**

Not seen in these open loop responses is the additional delay experienced in closed loop operation from the time it takes for the output signal to pass through deadband (backlash), threshold sensitivity (stiction), and resolution limits of the final control element. These open loop tests use a step change in output larger than these limits. In closed loop operation, a step change in output can occur for a step change in setpoint, but subsequent closed loop action to recover from overshoot or disturbances consists of gradual changes in the output. The additional deadtime from backlash, stiction, and resolution limits can be approximated as the limit divided by the rate of change of the signal (e.g., valve deadband divided by the rate of change of the PID output).

The open loop time constant (τ_0) is the largest time constant plus any portion of smaller time constants not taken as effective deadtime. While often called the process time constant, the largest time constant can occur anywhere in the loop. For liquid pressure and flow loops, the largest time constant is usually somewhere in the automation system (e.g., valve, sensor, transmitter, or DCS). Ideally, the largest time constant is the primary time constant of the process downstream of where the disturbances and manipulated flow enter the process as shown in Figure D-4. The effect of such a primary process time constant is beneficial. The ultimate limit of the peak and integrated errors are inversely to this primary time constant. The primary process time constant slows down an excursion resulting from a disturbance, giving time for the PID to catch up with it. If the largest time constant is in the measurement, the trend chart oscillation may look better because the amplitude is attenuated by the filtering effect of the measurement time constant. In an open loop test, you cannot discern the location of the largest time constant.

The open loop gain (K_0) is the product of the final control element, process, and measurement gain. Consider a loop with a control valve. The final control element gain (the change in flow divided by the change in % PID output) is the slope of the valve's installed characteristic curve. The process gain (the change in process variable divided by the change in valve flow) is the slope of a plot of the process variable versus valve flow. The measurement gain (the change in % PID input divided by the change in process variable) is 100% divided by the measurement span. **Because the PID algorithm uses percent signals, calculations of the open loop gain must use percent signals despite the fact that the PID block and graphics show the PID process variable, and in some DCS the PID output, in engineering units.**

Self-regulating processes are defined by a total loop deadtime, an open loop time constant, and an open loop gain called a steady gain (Figure D-1). A secondary time constant may be used to describe the initial lag (bend) in the response. The open loop gain is a steady-state gain that is the percent change in PID input from its initial to its final value divided by the percent change in PID output, giving a dimensionless gain. Liquid pressure and flow loops have a self-regulating process response. Continuous composition, pH, and temperature loops have a self-regulating process response, but the process time constant for large, well-mixed vessels is so large that in the time frame of interest for the PID (two to four deadtimes), the response resembles the ramp of an integrating process. For tuning and analysis, it is useful to treat self-regulating processes with a time constant order(s) of magnitude larger than the total loop deadtime as "near integrating" (Tip # 89).

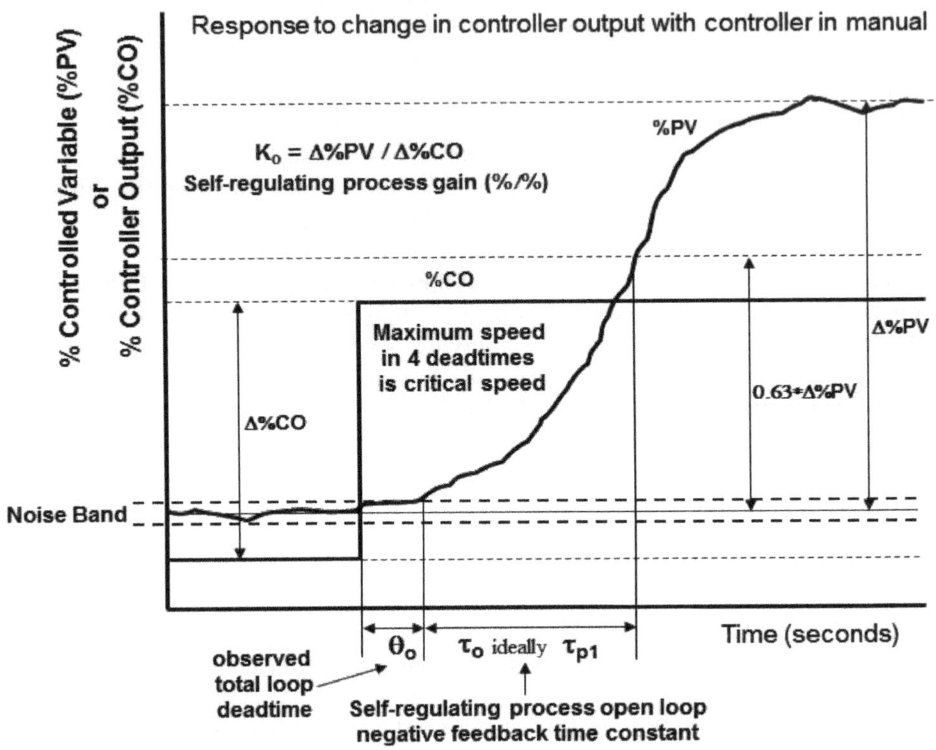

Figure D-1: Open loop self-regulating processes decelerate and line out at a new operating point (steady state) below a physical limit or Safety Instrumentation System (SIS) or relief device setting.

Integrating processes are defined by a total loop deadtime and an integrating process gain (Figure D-2). A secondary time constant may be used to describe the initial lag (bend) in the response. The integrating process gain is the ramp rate in percent of PID input per second divided by the percent change in PID output, giving a gain of reciprocal time units (1/sec). Level is the most common loop with an integrating response. Gas pressure has a near integrating or true integrating response, depending on the size of the pressure drop across the manipulated valve relative to the changes in pressure during an open loop test. Batch composition, pH, and temperature loops essentially have an integrating response unless it is altered by a reaction. An integrating process will exhibit a self-regulating closed loop response with a proportional-only controller. The distance of the new from the initial operating point decreases as the PID gain increases. Integrating and near integrating processes require aggressive proportional action. The steady-state gain divided by the open loop time constant of a near integrating process is effectively an integrating process gain. The maximum PID gain is inversely proportional to the process time constant or the integrating process gain. **Most integrating processes are so slow (integrating gain is so small) and the deadtime is so relatively small that the maximum PID gain is so large that the primary limit to how high you set this gain is user knowledge and noise.**

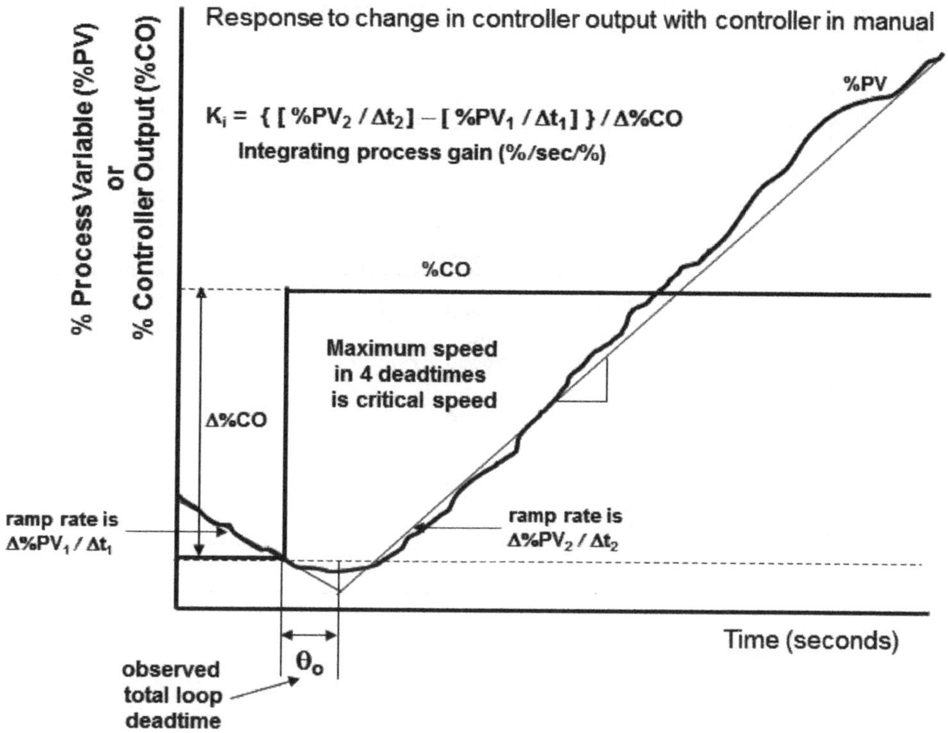

Figure D-2: Open loop integrating processes continue to ramp until hitting a physical limit or triggering the activation of a Safety Instrumentation System (SIS) or relief device

Runaway processes are defined by a total loop deadtime, a positive feedback open loop time constant, and an open loop gain called a steady gain (Figure D-3). The positive feedback time constant causes the continual acceleration. A secondary time constant may be used to describe the initial lag (bend) in the response. The open loop gain is a steady-state gain that is the percent change in PID input from its initial to its final value divided by the percent change in PID output, giving a dimensionless gain. Open loop tests are rarely done on runaway processes because of the safety concerns stemming from the acceleration. Manual control of a runaway process is extremely difficult. **Most true runaway processes are always operated in automatic or a higher mode, such as cascade or remote cascade. Tuning tests are done with the PID loops in automatic.** The large controller gains and integral times used for these loops provide a step change in controller output from the proportional mode and negligible ramping from the integral mode in four deadtimes. Polymerization and specialty chemical reactors that experience a rate of heat release from an exothermic reaction that exceeds the cooling rate can develop a runaway response to an increase in temperature.

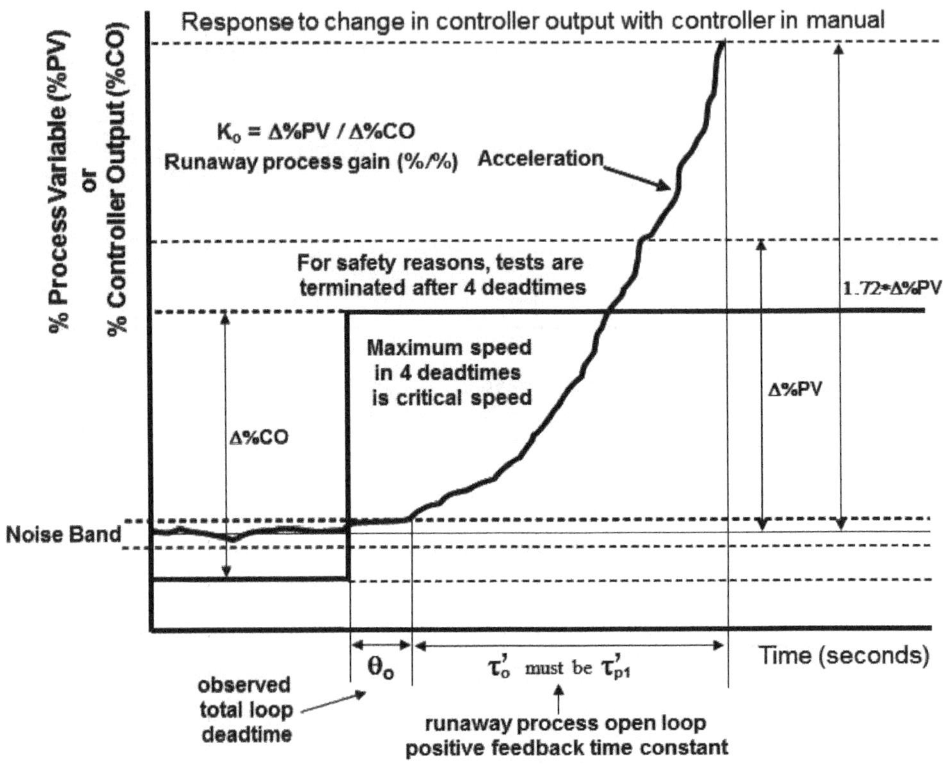

Figure D-3: Runaway processes continue to accelerate until hitting a physical limit or triggering the activation of a Safety Instrumentation System (SIS) or relief device.

A shortcut method (Tip # 89) can easily and rapidly identify the dynamics for slow self-regulating, integrating, and runaway responses because these processes exhibit a similar response in the first four deadtimes. The method is consistent with the action of a well-tuned PID that responds to and corrects an excursion from a disturbance in two to four deadtimes. A shortcut method for response analysis and tuning identifies just the deadtime and the maximum ramp rate during the next two to four deadtime intervals. By choosing the maximum ramp rate, the effect of a secondary time constant may be ignored. The %/sec ramp rate divided by the percent change in PID output is the integrating or near integrating gain (1/sec). The test can be done with the PID in automatic if the structure includes the proportional mode on error and the controller gain is high enough to provide a significant step change in PID output. The parameters can generally be identified in less than five deadtime intervals, making the test much faster for self-regulating processes that require four to five time constants to be able to recognize the steady state. The fact that the test can be done in automatic, is fast, and that the parameters are easy to identify makes it attractive for routinely checking dynamics and tuning whenever there is step change in the controller output.

The logic and computation for identifying the dynamics with the shortcut method are elegantly concise. When there is a significant step change in PID output, the deadtime is identified as the time till the point at which the PV either reverses direction or changes faster in the direction consistent with direction of the change in PID output, valve action, and process action. The process does not need to be at steady state. The PV is sent to a deadtime block to create an old PV, which is then subtracted from the new PV to get a delta PV. Every execution of the deadtime block provides a new delta PV. The largest delta PV over four deadtimes is divided by the deadtime and change in PID output. The result is the integrating process gain (%/sec/% or 1/sec). The PV and output must be in %.

The block diagram (Figure D-4) shows the source of the dynamics depicted in Figures D-1 through D-3.

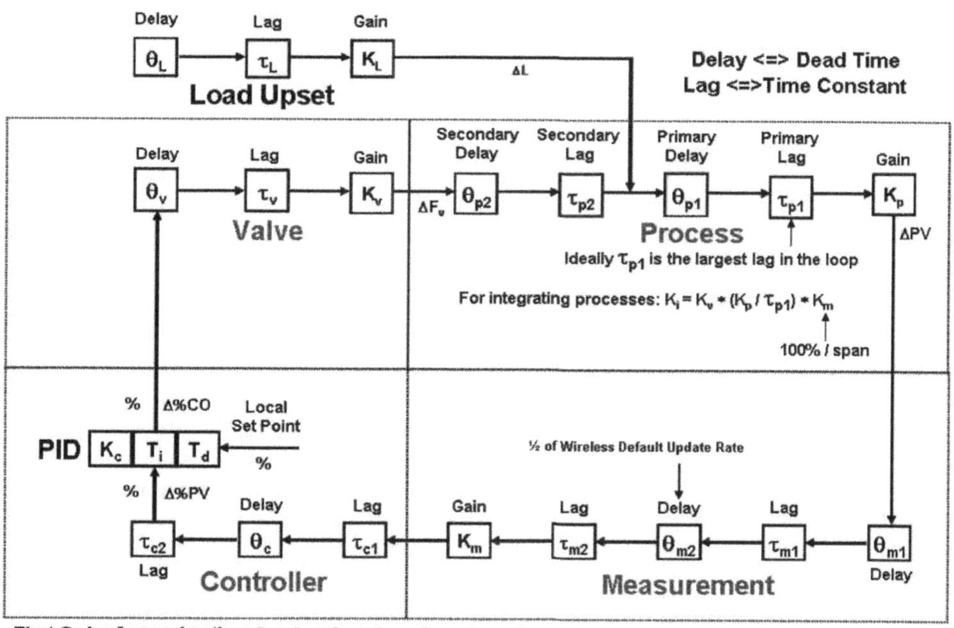

First Order Approximation: $\theta_o \cong \theta_v + \theta_{p1} + \theta_{p2} + \theta_{m1} + \theta_{m2} + \theta_c + \tau_v + \tau_{p2} + \tau_{m1} + \tau_{m2} + \tau_{c1} + \tau_{c2}$
(set by automation system design for flow, pressure, level, speed, surge, and static mixer pH control)

Figure D-4: The open loop gain is the product of the valve, process, and measurement gains, the loop deadtime is the sum of all the delays and small lags, and the open loop time constant is the largest lag.

Appendix E: Enhanced PID

The enhanced PID was developed for wireless but has proven to be useful for dealing with many discontinuous signals, particularly when the update times are large and variable. The discontinuous nature of the signal can be caused by problems and cyclic measurements, such as a temporary loss in communication, control valve backlash and stiction (Tip # 97), coated pH electrodes (Tip # 86), automatically retractable insertion pH assemblies (Tip #88), and at-line and offline analyzers (Tip #100).

Wireless measurement devices have a "default update rate" (time interval for periodic reporting) and a "trigger level" (threshold sensitivity limit for exception reporting) that are set as large as possible to conserve battery life. In the traditional PID the integral mode will continue to ramp while the PID is waiting for an updated measurement from a wireless device. In addition, when an update is received the traditional PID considers the entire change to have occurred within the PID execution time interval (ΔT_x). If derivative mode is used, the rate of change of the measurement is the difference between the new and old measurement divided by the PID execution time interval. The result is a spike in the controller output.

An enhanced PID controller for wireless will execute the PID algorithm as fast as a traditional PID controller for wired devices. A change in setpoint, feedforward signal, and remote output is translated immediately (i.e., within the PID execution time interval) to a change in PID output. However, with enhanced PID, integral action does not make a change in the output until there is an update. When an update occurs, the elapsed time between the updates is used in an exponential calculation that mimics the action of the filter block in the positive feedback implementation of integral action. If derivative action is used, the elapsed time, rather than the PID execution time interval, is used to calculate the rate of change of the process variable. The integral and derivative calculations are executed only once upon a change in setpoint or measurement. A threshold sensitivity setting is used when the PID output is within a normal range to prevent the occurrence of an update from noise. Figure E-1 compares simplified block diagrams of the traditional PID and the enhanced PID.

The traditional PID must be detuned to prevent instability if there is a long time between updates. The enhanced PID will continue to be stable for even the longest update time interval. For a measurement update time interval larger than the 98% process response time, the enhanced PID controller gain can be set equal to the inverse of the open loop gain (the product of valve, process, and measurement gain) to provide a complete correction in one execution for a setpoint change or update. The enhanced PID can suppress oscillations from a wide variety of sources. This reduction in variability results from the suspension of integral action and the wait in feedback correction until there is a more complete response. To achieve these benefits, the user must simply enable the enhanced PID option in the PID block, which automatically enables the dynamic reset limit option. No retuning is necessary to achieve a smooth response, but if the update time is larger than the process response time, the enhanced PID can be tuned with a gain that approaches the inverse of the open loop gain.

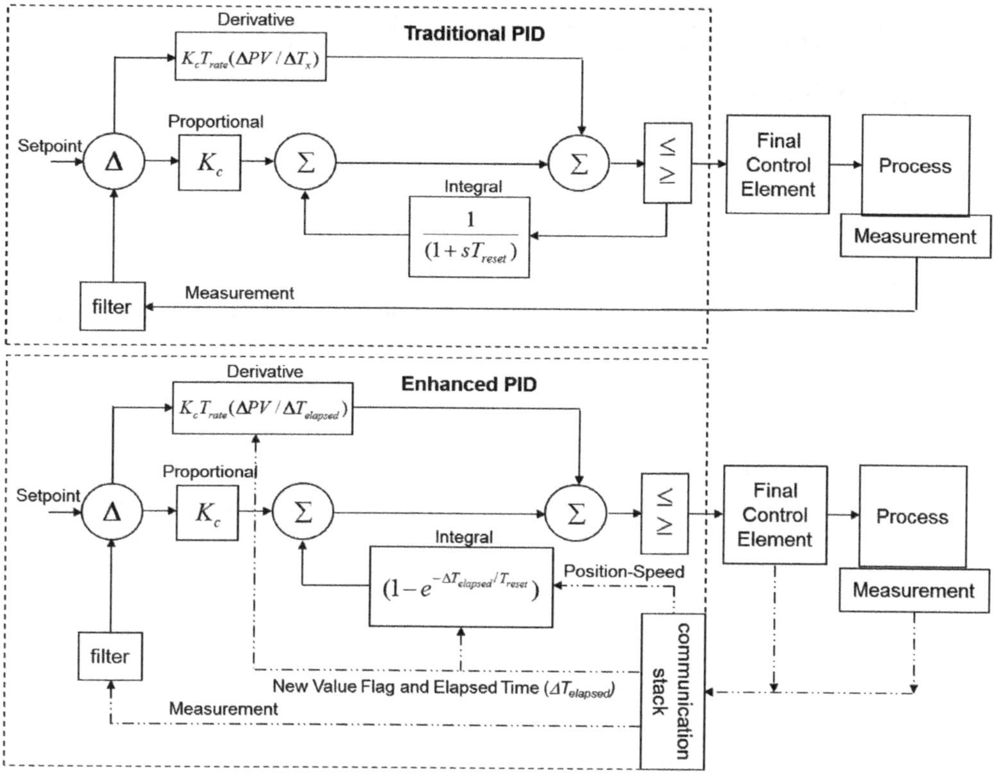

Figure E-1: Enhancements of PID for Wireless Prevent the Ramping from Integral Action and the Spikes from Derivative Action Resulting from Discontinuous Updates to Traditional PID

Appendix F: First Principle Process Gains, Deadtimes, and Time Constants

There are three types of processes; self-regulating, integrating, and runaway as shown in Figures F-1, F-2, and F-3, respectively. A self-regulating process will decelerate to a new steady state. An integrating process will continually ramp. A runaway process will accelerate until hitting a relief or interlock setting.

Over 90% of the processes are self-regulating. However, many of the continuous and fed-batch processes in the chemical industry with the greatest direct economic benefits behave and can be best treated as "near integrating" processes. The classic integrating process is a pure batch or level process. Less than 1% of the processes are runaway. When these exist, understanding the runaway response is critical in terms of safety and control because of the propensity to accelerate and reach a point of no return. Runaway responses are almost exclusively associated with highly exothermic reactors used in plastics and specialty chemical production.

This note develops the equations for process dynamics for back mixed volumes and plug flow volumes. The back mixed volumes section applies to volumes whenever an agitator pumping rate, an eductor or recirculation liquid flow rate, or gas evolution or sparge rate produces enough turbulence and back mixing to make the mixture more uniform in the axial besides the radial direction of the volume. Gas volumes generally have enough turbulence and a fast enough gas dispersion rates to be treated as a back mixed volume. The equations therefore hold relatively well for an evaporator and a single distillation stage due to turbulence from the vapor flow. The plug volume section applies to static mixers, pipelines, coil inlets, and jacket inlets where the turbulence from pipe fittings or internal mixing elements creates enough radial mixing to make the mixture uniform over the cross-section of the pipe or nozzle inlet but little axial mixing.

Back Mixed Volumes

For a back mixed volume, the process gains and time constants can be readily identified if the ordinary differential equations for the rate of accumulation of energy or material in the volume are set up so that the process output of interest (Y) is on right side with a unity coefficient. From this simple generic form we can identify the process time constant (τ_p) as the coefficient of derivative of the process output (dY/dt) and the process gain (K_p) as the coefficient of the process input (X). The process output (Y) and input (X) can be viewed as the controlled and manipulated variables, respectively. Many other terms can exist but these are not shown in the following equations. These missing terms can be categorized as disturbances.

If the sign of the unity coefficient of the process output on the right side is negative (Equation F-1a), the process has negative feedback. As the process output changes, the negative

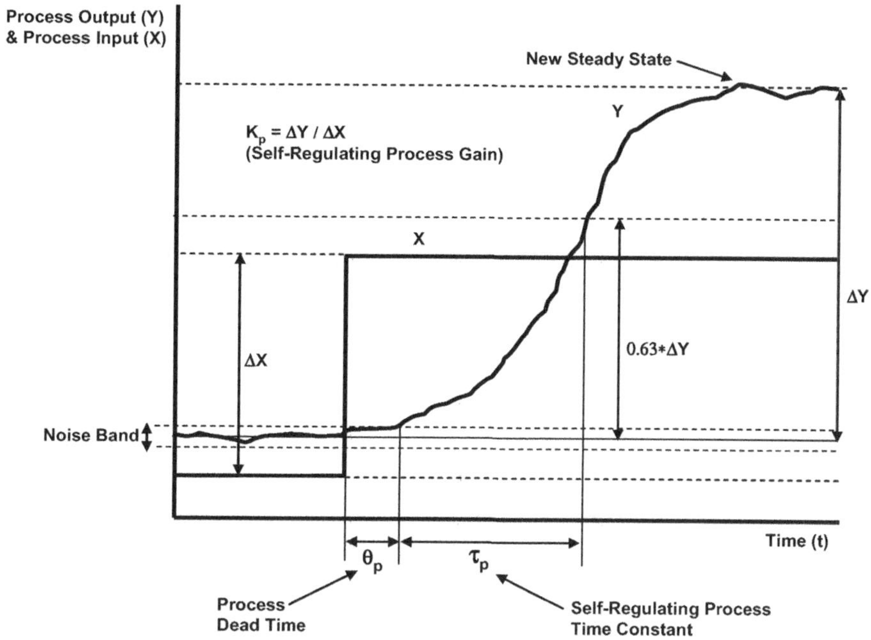

Figure F-1. Self-regulating (negative feedback) process

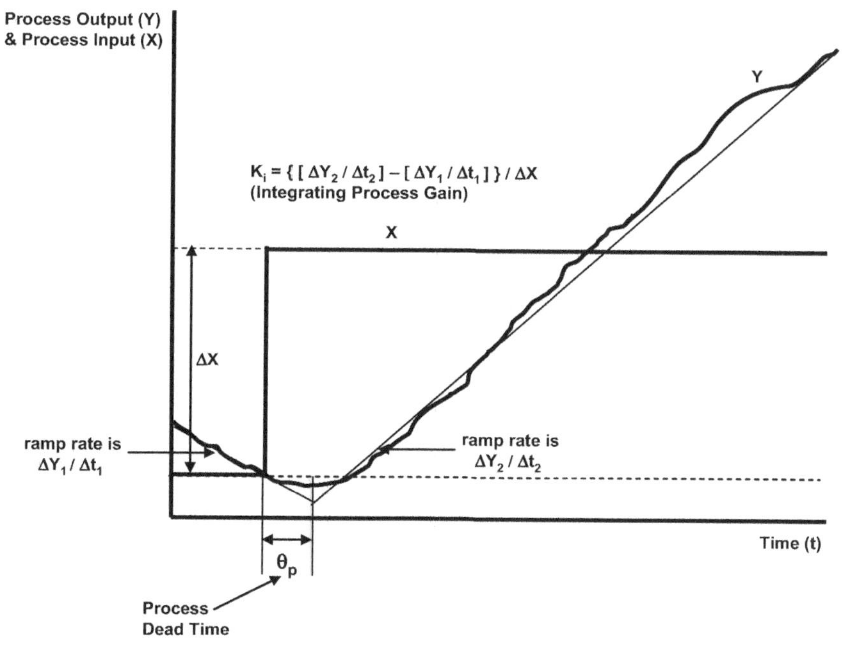

Figure F-2. Integrating (zero feedback) process

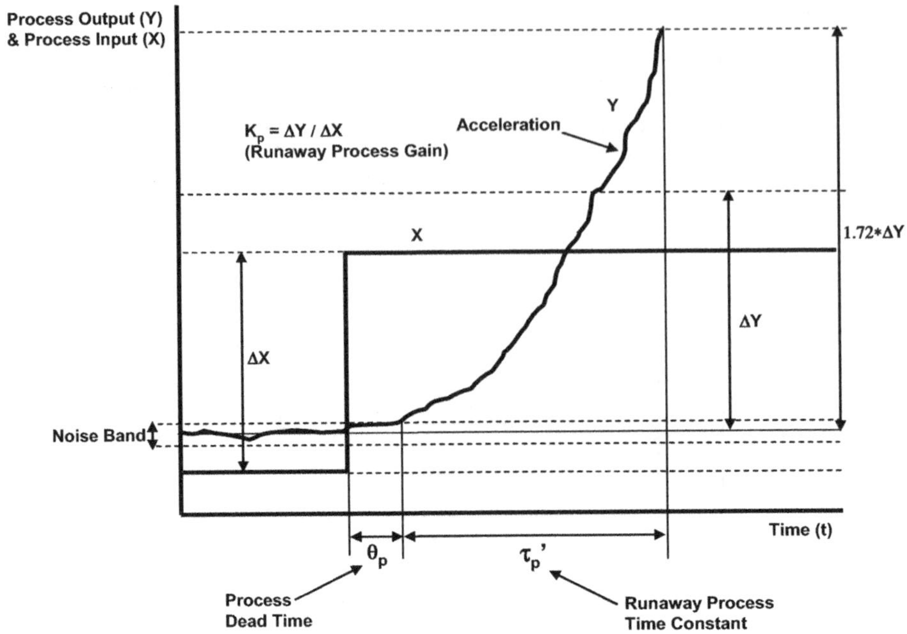

Figure F-3. Runaway (positive feedback) process

feedback slows down and eventually halts the excursion of the process output at its new steady state when it balances out the effect of the process input and the disturbances.

$$\tau_p * dY / dt = K_p * X - Y \qquad \text{(F-1a)}$$

The integration of this equation provides the time response of a change in the process output (ΔY) for a change in the process input (ΔX).

$$\Delta Y = K_p * (1 - e^{-t/\tau_p}) * \Delta X \qquad \text{(F-1b)}$$

If the process output does not appear on the right side (Equation F-2a), there is no process feedback. As the process output changes, there is no feedback to slow it down or speed it up so it continues to ramp. There is no steady state. The ramping will only stop when X is zero or balances out the disturbances.

$$dY / dt = K_i * X \qquad \text{(F-2a)}$$

$$\Delta Y = K_i * \Delta t * \Delta X \qquad \text{(F-2b)}$$

Often in the more important loops for concentration, pressure, and temperature control of large volumes, the time constant in Equation F-1a is so large that the time to reach steady state is beyond the time frame of interest. Since these loops with small deadtime to time constant ratios should be tuned with small Lambda factors (high controller gains) per Advanced Application Note 3, the controller only sees the first part of the excursion before the inflection point and deceleration by negative process feedback. In this case we have a "near Integrator" and Equation F-1a is best visualized as Equation F-2a with an integrator gain calculated per Equation F-2c.

$$K_i = K_p / \tau_p \tag{F-2c}$$

If the sign of the unity coefficient of the process output on the right side is positive (Equation F-3a), the process has positive feedback. As the process output changes, the positive feedback speeds up the excursion unless disturbances counteract the effect of the process input and output.

$$\tau_p * dY / dt = K_p * X + Y \tag{F-3a}$$

$$\Delta Y = K_p * (e^{t/\tau_p} - 1) * \Delta X \tag{F-3b}$$

Consider a mixed volume with a jacket and vapor space as shown in Figure F-4. There are liquid reactant feeds, gas feeds (sparged through the liquid and added directly to the vapor space), an outlet liquid flow, a vent gas flow, and a jacket coolant flow. There is normally multiple components interest. For example, consider liquid or gas acid and base reagent or reactant components (*a, b*) to produce primary and secondary liquid or gas products (*c, d, e*). Consider also there are typically water and nitrogen gas components (*w, n*).

The ordinary differential equation for the accumulation of liquid mass as shown in Equation F-4a includes inlet flows added directly to the liquid volume (ΣF_i), vapor flow rates from evaporation and vaporization (ΣF_v), and an outlet liquid flow rate (F_o). The liquid level depends upon density and cross-section area of the liquid. Equation F-4a can then be reformulated to Equation F-4b to include the process variable of interest, liquid level (L_l), in the derivative.

$$dM_o / dt = \Sigma F_i - \Sigma F_v - F_o \tag{F-4a}$$

$$d(\rho_o * A_o * L_o) / dt = \Sigma F_i - \Sigma F_v - F_o \tag{F-4b}$$

If we consider the density (ρ_o) to be a weak function of composition and therefore constant like the cross-sectional area (A_o) we can take these terms outside the derivative and divide through to get an equation for level (L_o) in the form of F-2a. Now it is clearly evident that the integrating process gain (K_i) for manipulation of flows in or flow out is simply the inverse of the product of the liquid density and cross-section area (Equation F-4d).

$$dL_o / dt = [1 / (\rho_o * A_o)] * [\Sigma F_i - \Sigma F_v - F_o] \tag{F-4c}$$

$$K_i = 1 / (\rho_o * A_o) \tag{F-4d}$$

The ordinary differential equation for the accumulation of gas mass as shown in Equation F-5a includes inlet flows added directly to the gas volume (ΣF_i), vapor flow rates from gas sparging, evolution, and vaporization (ΣF_v), and an exit gas flow rate (F_g). Equations of state such as the ideal gas law can be used to express this relationship for a given composition. Equation F-5a can then be reformulated to Equation F-5b to include the process variable of interest, gas pressure (P_g), in the derivative.

$$dM_g / dt = \Sigma F_i + \Sigma F_v - F_g \tag{F-5a}$$

$$d[(P_g * V_g)/(R_g * T_g)]/dt = \Sigma F_i + \Sigma F_v - F_g \tag{F-5b}$$

If we consider changes in the gas volume (V_g) and gas temperature (T_g) to be much slower than changes in the gas pressure (P_g) and therefore relatively constant during the integration step we can take these terms outside the derivative and divide through to get an equation for pressure in the form of Equation F-2a. Now it is clearly evident that the integrating process gain (K_i) for manipulation of flows in or flow out is simply the product of the universal gas coefficient (R_g) and the absolute gas temperature divided by the gas volume (Equation F-5d). This assumes a change in pressure does not significantly change the gas glow out of the volume, which is normally the case for a pressure drop across the vent valve that is large or critical.

$$dP_g/dt = [(R_g * T_g)/V_g] * [\Sigma F_i + \Sigma F_v - F_g] \tag{F-5c}$$

$$K_i = [(R_g * T_g)/V_g] \tag{F-5d}$$

The ordinary differential equation for the accumulation of energy as shown in Equation F-6a includes the effects of feed temperature, heat of reaction as a function of temperature, heat of vaporization, and heat transfer to the jacket. If we consider the specific heat capacity relatively constant and use the multiplicative rule of integration, we can express the differential equation in the generic form of Equation F-2a in terms of temperature to show the process feedback. The relative magnitude of the terms in the denominator of Equation F-6g determines the feedback sign.

$$dQ_o / dt = C_p * \Sigma(F_i * T_i) - C_p * F_o * T_o + (\Delta Q_r / \Delta T_o) * T_o - H_v * F_v - U * A * (T_o - T_j) \tag{F-6a}$$

$$dQ_o / dt = d(C_p * M_o * T_o)/dt = C_p * (dM_o / dt) * T_o + C_p * M_o * (dT_o / dt) \tag{F-6b}$$

$$F_f = \Sigma F_i \tag{F-6c}$$

$$T_f = \Sigma(F_i * T_i) / \Sigma F_i \tag{F-6d}$$

$$C_p * (dM_o / dt) * T_o = C_p * (F_f - F_o) * T_o \tag{F-6e}$$

$$C_p * M_o * (dT_o/dt) = C_p * F_f * T_f + H_x * R_x - H_v * F_v + U * A * T_j - [C_p * F_f + \Delta Q_r / \Delta T_o + U * A] * T_o \tag{F-6f}$$

For the manipulation of jacket temperature to control outlet temperature, the main process time constant (τ_p) is (positive feedback if heat of feed and reaction exceeds product of heat transfer coefficient and area):

$$\tau_p = (C_p * M_o)/[C_p * F_f + \Delta Q_r/\Delta T_o + U * A] \tag{F-6g}$$

For the manipulation of jacket temperature to control outlet temperature, the process gain (K_p) is:

$$K_p = (U * A)/[C_p * F_f + \Delta Q_r/\Delta T_o + U * A] \tag{F-6h}$$

For the manipulation of jacket temperature to control outlet temperature, the near integrator gain (K_i) is:

$$K_i = (U * A)/(C_p * M_o) \tag{F-6i}$$

For the manipulation of feed temperature to control outlet temperature, the process gain (K_p) is:

$$K_p = (C_p * F_f)/[C_p * F_f + \Delta Q_r/\Delta T_o + U * A] \tag{F-6j}$$

For the manipulation of feed flow to control outlet temperature, the process gain (K_p) is:

$$K_p = (C_p * T_f)/[C_p * F_f + \Delta Q_r/\Delta T_o + U * A] \tag{F-6k}$$

For manipulation of jacket temperature, the additional small secondary process time constant associated with the heat capacity and mass of the jacket wall is:

$$\tau_p = (C_w * M_w)/[U * A] \tag{F-6l}$$

Any change in the temperature at the heat transfer surfaces or the feed inlet must be dispersed and back mixed into the volume. This process deadtime (θ_p) is the turn over time that can be approximated as the liquid inventory divided by the summation of the feed flow rate (F_f), agitator pumping rate (F_a), recirculation flow rate (F_r), and vapor evolution rate or vapor bubble rate (F_v). Since this turn over time is computed in terms of volumetric flow rates, the liquid mass and the mass flow rates are divided by their respective densities as shown in Equation F-6m.

$$\theta_p = (M_o/\rho_o)/[(F_f + F_a + F_r)/\rho_o + F_v/\rho_v] \tag{F-6m}$$

If there is an injector (dip tube or sparger ring) volume, a change in composition at the nozzle must propagate by plug flow to the discharge points of the dip tube or sparger ring. The deadtime for a feed flow (F_1) is the injector volume (V_1) divided by the injection mass flow (F_1) divided by its respective density (ρ_1).

$$\theta_p = V_1/(F_1/\rho_1) \tag{F-6n}$$

The ordinary differential equation for the accumulation of liquid reactant mass (M_A) as shown in Equation F-7a includes the effects of feeds (F_i) with a reactant mass fraction (X_{Ai}), reaction rate (R_x), and outlet flow (F_o). The feeds can be from raw material, intermediate products, recycle streams, or multi-stage reactors. If we use the multiplicative rule of integration, we can express the differential equation in the generic form of Equation F-2a in terms of concentration to show the process feedback.

$$dM_A / dt = \Sigma(F_i * X_{Ai}) - (R_x + F_o) * X_{Ao} \qquad \text{(F-7a)}$$

$$dM_A / dt = d(M_o * X_{Ao}) / dt = (dM_o / dt) * X_{Ao} + M_o * (dX_{Ao} / dt) \qquad \text{(F-7b)}$$

$$F_f = \Sigma F_i \qquad \text{(F-7c)}$$

$$X_{Af} = \Sigma(F_i * X_{Ai}) / \Sigma F_i \qquad \text{(F-7d)}$$

$$(dM_o / dt) * X_{Ao} = (F_f - F_o) * X_{Ao} \qquad \text{(F-7e)}$$

$$M_o * (dX_{Ao} / dt) = F_f * X_{Af} - (R_x + F_f) * X_{Ao} \qquad \text{(F-7f)}$$

For the manipulation of feed flow to control reactant concentration (X_{Ao}), the main process time constant (τ_p) is:

$$\tau_p = M_o / (R_x + F_f) \qquad \text{(F-7g)}$$

For the manipulation of feed flow to control reactant concentration (X_{Ao}), the process gain (K_p) is:

$$K_p = X_{Af} / (R_x + F_f) \qquad \text{(F-7h)}$$

For the manipulation of feed flow to control reactant concentration, the near integrator gain (K_i) is:

$$K_i = X_{Af} / M_o \qquad \text{(F-7i)}$$

For the manipulation of feed concentration to control reactant concentration, the process gain (K_p) is:

$$K_p = F_f / (R_x + F_f) \qquad \text{(F-7j)}$$

The process deadtimes from turnover time and from feed injection are the same as computed in the section for temperature control (Equations F-6m and F-6n).

Plug Flow Volumes

For plug flow volumes where different streams are being combined, the process gain for controlling the temperature (T_f) or composition (X_{Af}) of the mixture (often a feed to a downstream equipment) can be computed by taking the derivative of Equations F-6d and

F-7d with respect to the manipulated flow stream 1 (F_1) to give Equations F-8a and F-8b, respectively. In both cases, the process gain is inversely proportional to total flow (ΣF_i).

$$K_p = dT_f / dF_1 = T_1 / \Sigma F_i \qquad \text{(F-8a)}$$

$$K_p = dX_{Af} / dF_1 = X_{A1} / \Sigma F_i \qquad \text{(F-8b)}$$

The process deadtime for the manipulation of a flow for stream 1 (F_1) is the summation of the injection delay for steam 1 and the piping delay from the point of injection to the point of temperature or composition measurement. For plug flow the residence time, which is the second expression in Equation F-8c completely becomes deadtime.

$$\theta_p = V_1 / (F_1 / \rho_1) + V_p / \Sigma(F_i / \rho_i) \qquad \text{(F-8c)}$$

The process time constant is essentially zero for true plug flow. For a static mixer there is some back mixing, the residence time in Equation F-8c is split between a deadtime and time constant per Equations F-8d and F-8e.

$$\theta_p = V_1 / (F_1 / \rho_1) + x * V_p / \Sigma(F_i / \rho_i) \qquad \text{(F-8d)}$$

$$\tau_p = (1 - x) * V_p / \Sigma(F_i / \rho_i) \qquad \text{(F-8e)}$$

It is obvious from the above that both the process gain and deadtime are inversely proportional to total flow.

Controller Tuning

The implication of the results can be best seen if Lambda is set equal to the total loop deadtime (θ_o) resulting in Equation F-9a for the controller gain. If the open loop time constant (τ_o) is large compared to the deadtime, the ratio of the open loop gain (K_o) to the time constant is the "near integrator" gain (K_i) shown in Equation F-9b which is the controller gain touted by the Simplified Internal Model Control (SIMC) as providing the best disturbance rejection. This controller gain is ½ of the gain from the Ziegler Nichols reaction curve method.

$$K_c = 0.5 * \frac{\tau_o}{K_o * \theta_o} \qquad \text{(F-9a)}$$

$$K_c = 0.5 * \frac{1}{K_i * \theta_o} \qquad \text{(F-9b)}$$

The time constant (τ_o) in the numerator of Equation F-9a is the largest time constant in the loop wherever it occurs. Hopefully, the process is mixed well enough and the instrumentation is fast enough that the largest time constant is in the process ($\tau_o = \tau_p$) and not the automation system. A large time constant in the process slows down the disturbance and is desirable. A large time constant in the measurement and final element is detrimental because it slows down the ability of the controller to see and react to disturbance, respectively.

The open loop gain (K_o) in the denominator is dimensionless. The process gain is actually the product of the manipulated variable gain, the process gain (K_F), the gain of nonlinear process variables, and the controlled variable gain. For a loop that throttles a control valve, the manipulated variable gain is the slope of the valve's installed characteristic. For the primary loop of a cascade control system, the manipulated variable gain is the secondary loop set point span divided by 100%. The controlled variable gain is 100% divided by the process variable span. Thus, changes in calibration span affect the computed controller gain.

Finally, the deadtime (θ_o) in the denominator is really the total loop deadtime, which is summation of the process deadtime (θ_p) plus all the small time lags and delays in the loop. While the names open loop time constant (τ_o), open loop gain (K_o), and total loop deadtime (θ_o) for the parameters in Equation F-10a are more definitive, nearly all of the control literature uses the terms process time constant, process gain, and process gain indiscriminately.

Nomenclature

Process Parameters:

A_o = cross-sectional area of liquid level (m^2)

A = heat transfer surface area (m^2)

C_p = heat capacity of process (kJ/kg*°C)

C_w = heat capacity of wall of heat transfer surface (kJ/kg*°C)

F_a = agitator pumping rate (kg/sec)

F_f = total feed flow (kg/sec)

F_g = gas flow (kg/sec)

F_i = feed stream i flow (kg/sec)

F_o = vessel outlet flow (kg/sec)

F_r = recirculation flow (kg/sec)

F_v = vaporization rate (kg/sec)

H_v = heat of vaporization (kJ/kg)

H_x = heat of reaction (kJ/kg)

L_o = liquid level (m)

M_A = component A mass (kg)

M_g = gas mass (kg)

M_o = liquid mass (kg)

M_w = mass of wall of heat transfer surface (kg)

P_g = gas pressure (kPa)

T_f = total feed temperature (°C)

T_g = gas temperature (°C)

T_i = feed stream i temperature (°C)

T_o = vessel outlet temperature (°C)

t = time (sec)

Q_o = total heat of liquid (kJ)

Q_r = heat from reaction (kJ)

R_x = reaction rate (kg/sec)

R_g = universal constant for ideal gas law (kPa*m^3/°C)

ρ_g = gas density (kg/m^3)

ρ_i = stream i density (kg/m^3)

ρ_o = liquid density (kg/m^3)

ρ_v = density of vapor (kg/m^3)

U = overall heat transfer coefficient (kJ/m^2*°C)

V_g = gas volume (m^3)

V_i = injection (e.g. dip tube or sparger ring) volume (m^3)

V_p = piping volume (m^3)

x = fraction of volume that is plug flow

X_{Af} = total feed component A concentration (mass fraction)

X_{Ai} = feed stream i component A concentration (mass fraction)

X_{Ao} = vessel outlet component A concentration (mass fraction)

Generic Terms:

X = process input (manipulated variable) (eu)

Y = process output (controlled variable) (eu)

Dynamic Parameters:

K_c = PID controller gain (dimensionless)

K_i = integrating process gain (1/sec)

K_o = open loop gain (dimensionless)

K_p = process gain (eu/eu)

τ_p = process time constant (sec)

τ_o = open loop time constant (sec)

θ_p = process deadtime (sec)

θ_o = total loop deadtime (sec)

References

1) Boudreau, Michael, A., McMillan, Gregory K., and Wilson, Grant E., "Maximizing PAT Benefits from Bioprocess Modeling and Control", Pharmaceutical technology IT Innovations Supplement, November, 2006.

2. Boudreau, Michael, A. and McMillan, Gregory K., *New Directions in Bioprocess Modeling and Control – Maximizing Process Analytical Technology Benefits,* Instrumentation, Automation, and Systems (ISA), 2006.

3. McMillan, Gregory, *Good Tuning – a Pocket Guide,* 2nd edition, Instrumentation, Automation, and Systems (ISA), 2005.

4. McMillan, Gregory and Cameron, Robert, *Models Unleashed – Virtual Plant and Model Predictive Control Applications,* Instrumentation, Automation, and Systems (ISA), 2004.

5. Blevins, Terrence L., McMillan, Gregory K., Wojsznis, Willy K., and Brown, Michael W., *Advanced Control Unleashed – Plant Performance Management for Optimum Benefits,* Instrumentation, Automation, and Systems (ISA), 2003.

Printed and bound by CPI Group (UK) Ltd, Croydon, CR0 4YY

13/09/2026

14961057-0001